Earth Remodeling Plans for Tomorrow's Home

— Possibly Eternal Peace of East Asia
in the Global Warming Era—

Yohji Esashi
(Honorary Professor of Tohoku Univ.
Former Consultant of FAO)

Obverse cover photo (Pamir Mountains) by Reo Ouchi

Earth Remodeling Plans for Tomorrow's Home

— Possibly Eternal Peace of East Asia in the Global Warming Era—

(未来のための地球再構築プラン　—地球温暖化時代における東アジアの恒久平和の可能性—)

Date of issue: Feb. 14 2017 (2017年2月14日　初版発行)

Author: Yohji Esashi (著者　江刺　洋司)

Publisher: Etsuo Ouchi (発行者　大内　悦男)

Publishing Office: Honnomori (発行所　本の森　984-0051 仙台市若林区新寺1丁目5-26-305)

1-5-26-305 Shintera,Wakabayashiku,Sendai 984-0051 JAPAN

Phone & Fax 022-293-1303 Email forest1@rose.ocn.ne.jp

URL http://honnomori-sendai.cool.coocan.jp

Printing Office: Haginosato Fukushikohjyo (萩の郷　福祉工場)

ISBN978-4-904184-87-5

Preface

At the end of 2014, COP20 (UN CLIMATE CHANGE CONFERENCE 2014) was held in Lima, the capital of Peru, and the allotment of greenhouse gas (GHG) emission by country was discussed. Although no agreement was achieved between the developing countries and the developed countries, however, the conclusion was put off to the next meeting in the following fiscal year.

Even if an agreement is reached in a day, the agreement only delays the depletion of fossil fuels that are limited resources in the earth. Therefore, it is predicted that the earth would be finally surrounded by atmosphere with a large amount of GHG. Human beings do not know how soon fossil fuels will be completely depleted, but、I can imagine that, at least 1,000 years later, humans will have exhausted fossil fuels and will have to depend only on renewable natural energy sources. In this article, I want to examine how people can live peacefully together in East Asia while enjoying basic human rights.

Contents

1. Strain-Related Background of the Existing East Asia

When I survey the strained relations of East Asia historically, I cannot but admit that Japan, the first country in Asia attempted modernization, gave the pressure to the surrounding nations, because Japan was a narrow island country with no resources for necessitating the import of both food and energy. Above all, China and Korea are still wary of Japan, although Japan has become a peaceful country with a peace constitution after World War II, which was caused in conjunction with the rapid population expansion followed by the introduction of the Industrial Revolution.

Although Japan was defeated in World War II, Japan accomplished rapid economic growth afterwards and reached to around 2,000 to show GDP of the world in second place. However, the economic growth in Japan had begun to slow in the 1,990s, and in place of it, the economic growth in China and Korea had become remarkable. The economic growth produced new issues that cause strain in East Asia. One is environmental accidents with the global warming, and the other is the rapid economic development of China which overtook Japanese GDP and may do even better than America's GDP soon. These were related to each other, but were generated with Chinese expansionism, the Sino Centrism with vast population, which laid feelings of strain for the whole world, above all, for the Asia-Pacific region.

This modernization in China, with the rapid increase of population, had begun using eminent coal as energy resources, and gasoline in the spread of motorization. The exhaust pollution gases from the thermal power stations and the cars, including the dust named PM2.5 other than the GHG, give a healthy uneasiness to the Chinese, afterwards they come over to Korea and Japan on the prevailing westerlies and give both nations uneasiness. But a more serious situation is the reality to form an economic zone of the origin of China money with the overseas Chinese, based on the Chinese self-confidence that was able to acquire economic power second place in the world. The concrete example is a design of AIIB (Asian Infrastructure Investment Bank) establishment, announced in autumn of the last year with the headquarter in Beijing, in which China finances accounts for half and the investments to the remaining half appeals to other countries.

Unlike European countries and the case of Japan, however, the population expansion in China has continued being relaxed exclusively in the form of the migrant worker to the foreign countries. In most cases, the Chinese lived in a group due to language barrier and according to the instructions of the government, and as a result opened Chinatown in each place, in which most of them have come to build suitable wealth.

The Chinese government continued one child policy to suppress the population pressure, but still the population in China rose to the 1.14 billion in about 1990 (World Economic Outlook Database:2014), and reached 13.4 hundred million in about 2010 when Chinese GDP overtook Japan (IMF:2014). At this point in time, it is said that the total of the overseas Chinese scattered in the world reached from 20 million to 50 million, varying according to an investigation. Since the world population at the point in time was about 68.67 hundred million (UNFPA), the Chinese population of 1.39 billion reaches approximately one-fifth of the world population.

Under such a foreknowledge, the Singaporean Prime Minister, Mr. Lee Kuan Yew in 1991 proposed the World Chinese Entrepreneur Convention to be held once every second year, which has aimed for contribution of the network among the overseas Chinese and for economic development of the host country. On the other hand, in 2000 to be able to foresee that China's GDP exceeds Japan with second place of the world, the communist party government of China came to hold the Overseas Chinese World Conference for Promoting the Peaceful Reunification of China in purpose to cooperate with the Chinese people in the world and to prevent the independence of Taiwan. Approximately at the same time, furthermore, the Chinese government has begun to hold "The Conference for Friendship of Oversea Association" every year in Beijing from 2002, in order to make use of financial power of the overseas Chinese in their own country's economic growth.

Thus, the economic growth of China increased rapidly, and gave the confident Chinese who occupied 1/5 of the human population to influence the world economy as greater consumers. In fact, a large number of Chinese who became rich traveled abroad as a tourist, and Chinatown took on course vigor. As a result, many countries including Japan trusted a China's economy and even accepted the card settlement as China's money yuan like the US dollar. These recent trends may be a sign that yuan as well as dollar has become a key currency of the world economy. It is thought that above-mentioned AIIB was born in such expectations of China. Such an overconfidence of China draws an air defense identification line on the Pacific without any consensus among surrounding countries. Many fishing fleets which raised Chinese national flags, throng to the territorial waters of Japanese Ogasawara Islands of the East Pacific for capture of red coral even under global restriction. This capture is used only under the Japanese maintenance power and is not a desirable act.

In order to demand eternal peace in East Asia, there must be no feelings of strain between China with second place of the world in GDP and Japan in third place. Fortunately, the population expansion in Japan already had completely stopped and, on the contrary, had entered in the times to be asked for the working population. In China, on the other hand, one-child policy expressed an effect, and the population growth is going to peak soon. Even if population increase is over and a country area is very large, the population size of China is too much. Can China provide the resources for a population of 1.4 billion? Further question is how the population capacity of China's area changes with the progression of global warming.

2. The Industrial Revolution Changed the Survival Factors Dramatically

In Japan, there is an old proverb that food, clothing and shelter are necessary so that a human being lives. Certainly, Japan is a marine nation consisting of islands spreading out in the temperate zone. The current century is expressed in other words with a century of water, but the word of the water is left out of Japan. It is certain that fresh water is an indispensable factor to life for all living entities. However, the fresh water does not become extinct from the planet earth because it circulates in the atmosphere. This fact suggests that Japan is located under the splendid environment condition blessed with fresh water, so that as global warming becomes more serious, Japan may be used to the extreme rain or snowfall because the atmosphere includes a large quantity of steam for warming.

By recent analysis of DNA constitution, the Japanese have been found to be completed in the interracial mix of many Asian people. The first passenger to the Japanese Islands escaped from the cold Eurasian Continent at the north end, when Japan was not separated by Mamiya Strait. They were Mongolian and, with the proof, many Japanese babies have an inexperienced irregularity in Mongolian spot to the backside. The second group to the Japanese Islands arrived at Kyusyu Island from the south along Philippines, Taiwan, and Ryukyu on the Kuroshio Current. Ancient Japanese first habitants are called "Paleolithic People", which were excavated from remains. The former was a movement of the invasion of the glacial epoch of a race, whereas the latter was a race movement for oneself to avoid submergence at rise in sea level time on the occasion at the end of the glacial epoch. The relic people in the former became the Ainu race in Karafuto and Hokkaido, whereas a native from farther south settled in Taiwan.

The third one occurred after a glacial epoch was completely over and the Japanese islands were completely separated from the continent in the sea. With civilization people of the continent went ashore by ship in the Japanese islands. The main route was through the Korean Peninsula. Especially of importance, it is the existence of people who came directly from Mainland China, and conveyed a character of kanji and rice growing civilization in Japan. People from the continent have formed the rule layer of Japan later. The historical fact is demonstrated by the exhumation of the ship-shaped clay image. In Japanese history, this time seemed to correspond from Johmon Period to Yayoi Period.

Mutual interchange began from this Yayoi Period, but the people who settled down in Japan were good at navigation and came to go abroad more positively, being described with appearances like Japanese pirates "WAKOU" to a later Chinese history book. The interchange of Japan with China beginning in this time made the basis to support Japanese culture established by introducing Chinese character "kanji". I think that Japanese is the superior language in the world with the flexible structure that can incorporate every world language. Further, I want to emphasize that the Japanese is in a relation of the far-off relative which received DNA of many Asian people, and thus, the genetic diversity has been produced naturally in the Japanese. Though the body was not very big, the genetic diversity had become the base that the talent person who could show ability in each field was developed in this way, before the industrial revolution happened in the West.

Before the industrial revolution occurring in 1760 in the UK, moreover, the above-mentioned facts suggest that the Japanese had splendid communication means to accept innovations. Furthermore, Japan had begun education at the place that called itself a private elementary school of the Edo period (for 264 years of 1603-1867) before the industrial revolution. The subjects of teaching were the national language, the classical Chinese which became basic, and Dutch or English.

In 1853 when American whale captured ship captain Mr. Perry called at a port near Tokyo Bay, accordingly, Japan already had taken in fruit of the industrial revolution voluntarily. From this time, however, Japan noticed the importance of acceptance of the European and American advanced civilization all over the country, and judged the reform of the political system with need for national modernization. In other words, from that time, energy increased in a serious factor for survival in addition to foods, clothing and shelter, although the information means was not yet recognized to be so important. Sad thing, however, Japan is a narrow country where the blessing of fossil fuel resources that became the energy sources was not gotten. In that way, Japan had caused World War II for the purpose of energy later.

The estimated population of Japan in the last year of the Edo was estimated to be approximately 3 million, but it has increased rapidly as soon as the industrial modernization began in the new government under the Emperor system, and reached 120 million 100 years later. Because Japan was a narrow island, the first plan to weaken the population pressure was population injection to develop uncivilized islands in the country, Hokkaido, Kunashiri and Etorofu. Besides, Japanese government performed policy-like emigration to United States and Brazil, etc. Only the effort as the nation of Japan, however, Japan was not able to give employment opportunity and foods to the population that continued rapidly increasing. During such an effort, the Japanese came to follow European and American colonial policy. The Japanese knew that they were going to secure the rights and interests on the continent before long, and judged that an action to prevent their expectations would help relax population pressure in Japan. To make matters worse, Japan came to pursue the new world in the continent in this way, too. This situation had given a chance to participate in World War I for Japan later.

Fortunately, Japan won this fight, and further got the sovereignty in a very large area of the Chishima Island, Karafuto southern half, the Korean Peninsula, Manchurian and Taiwan in the process until the time point. To the background of such good luck, it may be said that most nations were already able to read and write their national language through education. And, such an expanse was able to relax stress from the population expansion, but it has brought vast petro energy demands for both movements for the rule and for the extensive defense. As a result, the action of Japan that demanded energy sources had become a direct cause of later World War II. Here is the Japanese modern history that the success of the industrial modernization followed a population explosion in the Japanese who increased foods, but Japan was not able to supply enough crude oil for the modernized industry and the increased population. Moreover, the crude oil was not only the most important energy source but also the primordial materials of various kinds of products that were absolutely necessary for civic life.

Not only the industrial modernization but also the agricultural modernization had almost completed in most countries of the world during the 20th century, but the kind of energy source was different among each country. The electricity was supplied by oil or coal in the countries that have plenty of fossil resources. For the production of electric power, however, Japan was not blessed with any fossil fuel, and could not but depends on the generation from hydroelectric power stations based on the abundant rain and snowfall. For example, 4 hydroelectric power stations were established between the small river "Hirosegawa" which flowed by my house with water flow at about 10 m^3/s and 3 m altitude difference, before World War I, but 2 inefficient ones had to be destroyed with the construction of the thermal power station which is operated with the heavy oil imported from foreign countries.

As for the defeat in World War II, many Japanese went back to Japan, and thus the population pressure in Japan had deteriorated, not only the energy self-sufficiency rate (about 4%) but also the food self-sufficient rate (40%) at once. Thus, the factors for Japanese in the Edo era were food, clothing and shelter, but the industrial evolution and the defeat had added energy to them. Moreover, one of the energy sources was crude oil from which organic matter, such as plastic, vinyl and medicine etc. were produced.

In addition, most humans were distributed over the temperate zone with the four seasons. They must take warmth in the winter season depending on the burning of biomass in old days, but depending on heat of fossil fuel lately. From the above-mentioned facts, it became clear that we must add energy to 3 elementary factors of foods, clothing and shelter. However, all energy sources were fossil fuel, excepting hydraulic power generation and, though extremely few, warm geothermal power generation, until the peaceful use of atomic energy began after World War II.

All kinds of fossil fuel exhaust CO_2, which absorbs the long wave light from the sun and the radiant heat from the earth, and surrounds the earth with a warm atmosphere. In addition to CO_2, nitrous oxide and N_2O are generated from ammonium nitrate of compound fertilizer, and methane(CH_4) called natural gas which cannot be ignored as warming gas, even if their contents are slightly less in amounts in the air, since the former and the latter absorb infrared rays from the sun 300 times more and 20-30 times more, respectively than CO_2. Above all, it is to say that CH_4 itself as well as oil and coal, the fossil fuels are to be limited resources reserved in the earth, and they are destined to deplete at some time soon. Moreover, it is very important that the consumption speed of the fossil fuel is high in proportion to the increase of the population, and that we could not leave the resources for various kinds of organic matter products.

Unlike communication with words, the information revolution in the 2000s offered many communication tools and may lead to the shift to the post-industrial society. Both the paper and the pen became unnecessary for communication. On the other hand, the all apparatus connecting to the computer, the cellular telephone and the robot etc. would not work where electricity is not available. All begins to operate only depending on electricity. Judging from the relations of both mentioned above, the communication is a small existence in comparison with energy. If energy supply is stopped, we cannot live so easily without communication.

I have ignored the importance of fresh water as an important factor to life, because Japan is blessed with rain and snow falls on the islands located in the Asian monsoon zone through the year, and it is the country where there is not dry season so far. In fact, however, human beings live with fresh water as well as with other living entities, except the microbes which use a substance beside H_2O as an electron donor. Plants having various pigments, mainly chlorophyll, break down $H_2O \rightarrow \frac{1}{2}O_2+2H^++2e$ using energy from the sun in the daytime, thus electron and hydrogen, thus the plants get abundantly bioenergy ATP and reduction power NADPH in them, and thus fixes neighboring CO_2 to carbohydrates in a chloroplast. This process is photosynthesis, but when night comes over as the sun hides behind a mountain, the plants begin to take electrons out of H_2O through a TCA cycle using the carbohydrates which are saved within cells at daytime as a substitute of light, and finally produce abundant ATP and NADPH through the oxygen respiratory system in which the electron is delivered to O_2 in the atmosphere, thus making various high molecular organic substances even at night.

The animals including humans live through a process that is basically similar to the process of a plant at night. That is, the animals live by eating the carbohydrates, lipids and proteins produced through the photosynthesis of plants, and these foods basically play a role to take out an electron from H_2O and to supply ATP abundantly through an electron transport system called oxygen breathing. A part of produced ATP is used for the reduction of NADP. When the bioenergy and reduction power of the quantity were thus obtained using the electron flow from water, the animals grew and descendants increased. This valid mechanism is common to all creatures, but if the natural environmental condition is different, the different food-chains bring about the characteristic ecosystem as Biodiversity Convention is applied to maintenance.

In addition, the water carries various kinds of materials that flow through the body as a liquid, and contributes to temperature control of the body. It is natural that the fresh water is indispensable to both farming and manufacturing. The Japanese natural environments were rich without letting the survival of human beings to be conscious about how fresh water was an important existence so as to leave only food, clothing and shelter for words to live. Therefore, I want to decide to list the fresh water and the food for human being survival with "fresh water + food" in combination from now on.

Until now, I have described 5 kinds of factors necessary for human survival, clothing, food and shelter in Japanese, "energy" after the industrial revolution, and "communication" after the information revolution. Up to the present, however, all information appliances do not work without some kind of energy supply. Of course, all of fresh water, food, clothing and shelter were not supplied without fresh water. If we do not have a house to sleep safely in, we cannot live. After fossil fuel dries up, the clothes will be made with any biomass. For such reasons, I want to display the elements for human survival as follows: energy > fresh water + food > shelter> communication > clothing.

3. Global Warming Responding Dramatically to Population Increase over the Capacity of Earth

It may be said that the failure in the modern history of Japan is not to understand the priority of mentioned factors for human survival. Since the education system of Japan enables Japanese to accept influences of Industrial Revolution, the industrial modernization in Japan was accomplished at almost the same speed as that in Europe and America. But the developed industry in Japan had caused the sudden population expansion, for which the Japanese government could not respond appropriately right away because Japan was narrow. Without immigrant policy, Japan had depended on the military authorities in imitation of Europe and America, and it had gone to demand the rights and interests on the continent. In this way, the oil had become more and more indispensable for Japan.

Europe and America grasped this Japanese weakness and came to restrict the export of oil to Japan to limit the expansion of armaments of Japan. Then, the military authorities raised a slogan of "Large Asia Prosperity Sphere" to the nation and came to demand oil resources from Indonesia which was a Netherland colony, thus dragged Japan into World War II. A sad thing, they had run about the war without enough recognition that there were not wells producing oil in large quantities in administrative authority areas of Japan. Therefore, the releasing down of the atomic bombs were not necessary for the end of the war, for the situations that even we children were sent to investigate the number of neighboring pine trees so children could feel a day of the war's defeat. The military authorities were said to have moved a tank, warship etc. using the pine oil which was extracted from the stocks of pine trees.

In the case of World War II, we Japanese gradually used the fossil fuels exhaustively, and finally were cornered to depend on only biomass. Now, human beings must notice it being in the identical situation. By this article, fossil fuels want to be recognized anew that these are limited resources heading towards depletion earlier. Then, even if we used up all the fossil fuels, we could cook rice, take a warm bath, and run a bus using the biomass which is called wood, because woody plants are the most important source of CO_2, which acts on the slowdown of global warming. A commensal place with nature which humans demand is the forest, and the trees which are the main members of the forest are not allowed to be cut down so easily.

On the other hand, the forest has been known as common sense terms that not only the trees stock CO_2 in a trunk, but also steam H_2O by transpiration into the atmosphere. The H_2O released from the forest as well as the H_2O released from the ocean in the atmosphere has become the source of the rainfall and snowfall. Therefore, as people live apart from the sea, people will come to depend on the blessing of the forest for drinking water rather than on the blessing of the sea. Thus, the people making much of the need of the maintenance of natural environments play an active part based on understanding of the structure of the nature, by making the political party that advocated green in some countries. Nevertheless, the world's largest forest of Amazon which fixes CO_2 of enormous quantity to the organic matters get into such a situation that may be cleared for cultivation of useful crops. In the dispatch, the United Nations had an opportunity to draw any wisdom that was in need to promote the use of the limited resources in recycling societies for people living in the future.

It was the first Earth Summit of 1992. The first free talking in the subcommittee for biological diversity treaty that I was invited and participated in about future population capacity of the earth. The values of each committee were greatly different by estimation, in which I spoke to 8 billion. In contrast, a certain committee described that it could increase to 14 billion possibly, but remarked that it would be weeded out by the outbreak of a certain infectious disease and decreased to an appropriate level. In the background of his estimation, I had supposed that there was an estimation of the biomass production capacity on the very large area of Siberia in the time when a mammoth lived in the primitive atmosphere which included past CO_2 in large quantities, and when some kinds of grasses as wheat grew only during a season when sunshine and moderate temperature were kept. In any case, Ebola, an infectious disease in the summer of last year, gave heaviness in his remark.

However, I could not understand his estimation. Surely, population capacity of the earth will be fixed with the potential supplying of foods. At that point in time, when most human beings will have exhausted available fossil fuels, it seems that the concentrations of CO_2 and CH_4 will have reached high levels considerably, the warming will extremely advance, the sea surface will rise, and the central area of the continent will turn to desert in a further wide area. Even if the utilization of natural energy and atomic energy is accelerated, the time of using up of the fossil fuels is only delayed, because of the limited resources of both uranium and thallium and of the un-predictable weather change in the future. Therefore, it seems that the area suitable for farming decreases on earth.

Needless to say, the decisive factor to control a supply power of food is fresh water. In other words, the factor to decide a population capacity in each nation is a supply capability of fresh water in a corresponding country. However, the meaningful supplying capability of the fresh water seems to associate directly with both rainfall and snowfall unevenness through the year and in every area within the nation. That is, there is no fresh water supply despite a superior nation, even if rainfall or snowfall is seen only in the small portion in the nation, or even if a rainfall or snowfall is blessed only in a certain season.

I do not foresee what time, but humans will use up fossil fuels sometime later. At that time, it is certain that planet earth will become much warmer than now. In fact, the first abnormal weather associated with the warming had been reported year by year everywhere, particularly in the last year. In Japan alone, there had already been heavy snowfalls twice causing damages on the Coast of Japan Sea, although the heavy snowfall was generated in the mountain zone in an average year. This phenomenon was estimated to be a result from the rise in seawater temperature of the Japan Sea. In fact, Japan Meteorological Agency reports that the temperature of seawater at depth of the water 200m rose to 0.6℃ for 20 years from 1992 to 2012 by measuring at 3 points in the Japan Sea. It should be higher in the water temperature at the upper layer, and so it is thought that the atmospheric steam pressure on the Japan Sea area increases remarkably. If the prevailing westerly of periodic wind begins to blow, on the contrary, it seems that the atmosphere temperature above the Japan Sea begins to decrease rapidly by capturing the evaporation heat, and as a result, it brings heavy snowfall in a foreside of the Japan Sea. Such a weather change lets global warming to question people around the world.

Thus, the rises in the concentrations of CO_2, and CH_4 etc. in the atmosphere have been already accompanied by the rises in temperature in both marine water and atmosphere. The ocean occupies approximately 70% of the surface area of the earth, and its role supplying steam is far superior to the forest. The appearance of the very large desert opening from North Africa in the Middle East and of the dry land in North America, along the tropic of Cancer, and similarly the appearance of the deserts in Australia and South Africa and of the dryland in South America, along the tropic of Capricorn, have been explained meteorologically by an atmospheric flow. However, the very large deserts in the Gobi and Tarim Basin areas in China and the drying areas from North America to Canada spread out in the temperate zone, and their appearances are not explained in the meteorology. Only those places are distant from the ocean.

There is the story that the fossil of the dinosaur was found in the desert of this Chinese interior. Moreover, the stock trace of a fossilized tree is excavated in the remains of Loulan located in the northeast part of the Taklamakan Desert. According to these historical evidences, it is obvious that an ecosystem was established in this place of Chinese interior in ancient times. But, the Han race that set up the house on the ground soon was going to make a retaining wall afterwards to prevent the southing of the Mongolian race that held a residence in the northern ground. It was the Great Wall, a world heritage of China. The Great Wall was built with burning bricks at the beginning. Since the fuel used was cut wood, the forest deterioration began gradually, and it became hard to cut trees down. As the retaining wall of the long and extensive castle goes to the west end, the state is left as a picture collapsing for use of the sun-dried bricks. Because there was the forest in the Chinese interior as from these evidences, the steam circulated within the range of the atmosphere and showed that there was a food chain that was based on natural environments. In addition, the first global environment ravager in the history of humans might be the Chinese who took action to protect their own peaceful life.

4. Fresh Water is Infinite as Resources, while the Ground Water is Limited

The next destruction of the large-scale global environment in the China's continent inside was caused by the policy of the Chinese leader, Mr. Mao who was ignorant of science. Probably, I think that he did not even have the biological basic knowledge of high school level, which all creatures, except special ones, live depending on bioenergy acquired in an oxidation process of hydrogen (H^+) and electron dissociation of the water molecule (H_2O). All animals, including human beings, must take in food to take an electron out of water. In contrast, the plants live like an animal in the night using organic matter that has been produced by photosynthesis during the daytime, but can let water dissociation using sunlight in the day. In any case, they cannot live and grow if the suitable quantity of supply of fresh water is not promised.

Moreover, the higher plants must absorb a large quantity of fresh water through a rhizome system from the soil in order to lower leaf temperature by transpiration to photosynthesize organic substances. On the other hand, the animals cannot live only with a fluid volume within foods and require a large quantity of fresh water to drink. Even domestic animals grazing grasses, a large quantity of fresh drinking water is indispensable for them like human beings in winter season to assume dry grass bait.

Similarly, a good state as the industry decided on both the farming and the stock-raising as to how far the fresh water supply in the area becomes through the year. As a matter of course, the supply of fresh water in the area is determined in principle by the degree of the rainfall and snowfall in the area as well as the quantity of the underground water. However, the groundwater is a blessing of the rain and snowfall that poured into the neighborhood before. On the other hand, the rain and snow in the area is basically a present with the steam circulating within the range of the atmosphere evaporated from both the ocean and the forest. Therefore, the groundwater abundance in a certain area in the continent will be basically proportional to the degree of rainfall and snowfall.

An oasis to keep for the benefit of the water vein through the underground from the high mountains is no exception. And, the supplying amount of the groundwater appears with the degree of rain and snowfall and with a constant time lag which is a late relation with a groundwater level so that the width of a stratum is big. Thus, the supply of fresh water caused by the rain and snowfall may be thought as endless since the sea area accounts for about 70% at the earth surface, but the supplying power of the ground water may become limited temporarily. In the zone of the limited rain and snowfall, drying up of the groundwater becomes possible if human beings excessively use the groundwater.

In China, unfortunately, the atmosphere circulation of water from the forest has already become extinct with the construction of the Great Wall, suggesting that it is poor in the supply not only of rain and snowfall but also of the underground water in the Chinese continent interior. Unfortunately, the Chinese leader, Mr. Mao has hurried newly construction of China as he definitely could not understand the situation mentioned above. The policy that was advocated by him was the history that was well known as the Great Leap Forward from 1958 to 1960, in which he demanded a great increase in production in agriculture and industry.

However, he forced extremely stupid instructions on the people there. In agriculture, he organized people in the commune by which he aimed at the maximum yield of cereals per unit area of the farmland. Stupidly, the agricultural method recommended there was the technique called dense planting by high-density dissemination and then the personal inventive ideas had been completely ignored. This agricultural method may be applicable under the natural environments in the northern country such as Russia, because plants met the fresh water supply by lower evaporation rates and they were exposed to cold windows even in the summer. However, Mr. Mao was deeply attached to the imitation of Russia-born agricultural methods without being able to judge whether these methods were suitable for China where a fresh water supply power is poorer than Russia and temperatures in the daytime is higher.

Thus, the application of dense planting cultivation in the inlands of China with little rainfall strengthened transpiration explosively, invited the consumption of groundwater, and has annihilated, even the original ecosystem of farmland. The result of the imitation of these stupid agricultural methods has left the name for history as the Great Chinese Famine or Mao's Great Famine (1958-1961). Not only the farming by collectivization of agriculture had exhausted the farmland in the inland of China, but also it caused many to starve reaching 60 million people from 10 million people, although I do not know which document is accurate.

On the other hand, another purpose of the Great Leap Forward in 1958 was industrial promotion, and then the production of steel which various kinds of industrial materials was reinforced. The resources used for a blast furnace was fuel. The biomass was easy to obtain as well as the coal that China held as resources in large quantities. Early in the history of humans, China cut down trees in the forest to bake bricks, but China has again performed deforestation on a large scale to begin dissolving pig iron even in modern times when the importance of the work of the forest began to be recognized. In fact, the deforestation by China over this twice continued thrusting severe environmental situations for the Chinese now, but it is certain that they are the large scale destruction of the ecosystem particularly to the Eurasia midcontinent, like the reduction and extinction of the Aral Sea area due to excessive Russian immigration after World War II. Judging from these points, the destruction of natural environments caused by humans in the Eurasia midcontinent is not a negative inheritance that is permitted. As a human community, what we leave behind in human history to future generations who will live on this small planet earth matters.

5. Agreement of COP only Prolongs Human Life on Earth

When the forest was cleared once in a wide area, the reproduction of the forest becomes impossible if the area is not blessed with the rain through the year. If there are dry and wet seasons in a year, a period of the wet season is short, and there is little rainfall, it is usual to continue going to ruin as for the ecosystem derived from the forest in the area irreversibly. Usually, it is limited to the tropical rain forest zone where the resumption of nature that went to ruin is possible. Accordingly, it is an extremely natural thing that the ecosystem traces the course of the collapse in the area where the existing forest that serves as a source of supply of the steam was artificially destroyed intensively as in the China's interior. In such an area that had become weak artificially, even the reforestation by human planting is not easy unless people are greatly concerned with periodical sprinkling.

Certainly, the further dilapidation of the natural environments of the Chinese plateau in the 20th century is due to a policy to be said as the Great Leap Forward by the Chinese Communist Party government fascinated by Mr. Mao. Meanly, he further commanded the Cultural Revolution campaign from 1996 to conceal his own great misleading in 1958-1961, by which many cultural historic remains have been destroyed idly. However, the government of the Chinese Communist Party faced the desiccation to approach the capital Beijing and the suspension of water supply in the Yellow River. It was the first time when they noticed the importance of his misleading. Then, it seemed that the first proof of their reflection was the Forest Law established in 1979 after the end of the Cultural Revolution Movement. Though belated, the Chinese government has begun to press action on planting woody plants in addition to the management of the country by the conservation of forest sources.

However, it was too slow for them to notice. The atmospheric concentration of the global warming gases derived from fossil fuels began to already rise, and some scientists have begun concern the future global environment. In fact, the concentration of CO_2 that I learned at a junior high school was 310 ppm, but increased 330 ppm in the 1980s. At that time, nobody had been able to play in the small marsh where I had enjoyed bamboo skating in the days of a primary schoolchild. It was the times when I already sensed warming of the earth bodily and when the global warming began to be discussed about whether or not it depends on discharge of CO_2. If so, the evaporation quantity of water might become superior to the quantity of rain snowfall in more areas in China. As the progress of the warming increases the quantity of evaporation under the situation, planting may hardly succeed. When they noticed the importance of the planting, it has been the times in which they must use limited groundwater so that they succeed in planting. In many areas of planting in the middle country, there was the risk that might promote further desertification.

By the way, the substantial amount of discharged CO_2 should be shown with differences between the quantity of real discharge and the quantity of absorption fixation by the trees. In COP 20, all countries should cooperate by an augment of the emission credit of CO_2, since they are concerned with real atmospheric CO_2 density. However, although China and India are large countries of CO_2 discharge, they continue to take the situation of developing countries still in COP 20, and do not yet reach the agreement with developed countries in 2014. This opposition with the developed countries and developing countries continues has already continued during 17 years since COP3 for Kyoto Protocol to the United Nations Framework Convention Climate Change in 1997. China, which is a self-styled latter representative, gets GDP of the second place of the world during this period, but disappointingly it does not seem to feel the responsibility that had greatly reduced a CO_2 absorption source under the slogan named the Great Leap Forward. In the background, it seems that China may have the prospect of having enough sources for fresh water supply by which the forest can produce on its own. Finally, it important to make a clear prospect in absorption—fixing the system of the CO_2 for tomorrow's earth. Needless to say, the international talks in COP are necessary for the reduction of the emission of CO_2, CH_4 and N_2O, but we must do something to prevent further atmospheric accumulation of CO_2.

Saying that we must do something by all means, this is because CO_2 itself is toxic for all animals including human beings, in addition to a source of global warming. There was a volcanic disaster in the crater lake of Lac Nyos in Cameroon in 1986 that I want to remember. On the occasion, approximately 1,500 people died by exposure to volcanic CO_2. For oxygen breathing in animals, it goes without saying that a big difference in CO_2 density in the inside and outside of the alveolus is necessary for CO_2 exchange to send out CO_2 as waste of blood. According to the instructions from the Japanese Fire and Disaster Management Agency, it is written that we must control the CO_2 density out of the current atmosphere to several times to exhaust CO_2 in blood smoothly. If CO_2 density exceeds 50,000ppm, human beings with dyspnea, and the blood pressure rises, thus causing various kinds of severe symptoms. Current CO_2 has already exceeded 400 ppm, and so it will not take long to reach its critical CO_2 concentration that will cause further global warming and threaten the survival of our own.

Each country in COP will agree by effluent control of CO_2 sometimes soon. Even so, however, the fossil fuels continue being used and dry up sometime soon. At that time, the biggest problem is whether or not humans can hold down atmospheric CO_2 density to the level that the slightest adverse effects will not harm the life of animals.

I have known that the atmosphere at the time when the earth was born consisted mostly of CO_2. And the most which existed as CO_2 was returned to the earth after being fixed as limestone. But I do not know what percent of the ancient CO_2 become fossil fuels, and what percent of them becomes irreproducible organic matter such as the plastic returned to the earth either. In the case of the future fossil fuel drying up, the biggest problem thrust at us who live in the present age, whether humankind can control the atmospheric CO_2 density and keep it at least 1,000ppm as well as a weather change called global warming. It is not merely a problem to get off with only global warming.

To suppress the excessive rise in CO_2 density is important here. To that end, both the direct use of CO_2 as industrial material and its temporary retention will be considered. The latter is already pushed as development of the means to shut in CO_2 in the deep bottom of the sea and into the deep soil. However, the problems such as earthquake induction or the neutralization of the sea remain hard to predict. The best means to hold a rise in CO_2 density is to store CO_2 in wood and keep recycling alive in societies in the future.

6. Energy, Fresh Water + Food as Human Survival Condition

I have stated that the increase in CO_2 would include the crisis of human extinction. However, the land covers only about 30% of the earth, and the remaining is the ocean, which is the greatest source of fresh water supply. The area of lands must continue reducing while global warming is increasing. On the other hand, the ocean must continue acidifying with increasing CO_2 density, following the collapse of marine ecosystems. Then, I want to further to arrange the elements concerning human survival.

In the case of most, the estimation of the capacity of the world population has been done based on the estimation of the food supply. All types of food are organic matter, which are constituted from a carbon atom, but it overflows as CO_2 now within the atmosphere on every land on the earth. In addition, the carbon elements are returned to organic matter of foods if fresh water exists except for microorganisms, in which a water molecule H_2O is divided into a pair of H^+ and electron and O_2 by light (the sun) acts on organic reduction of the CO_2 there. Therefore, the population capacity of the earth in the future after the fossil fuel dries up will not be a failure of food supply, and it will be estimated at land area met with both the sunlight and the fresh water. By the way, there are the marine products such as fishery products, whales and seaweeds, too, but not the staple foods. However, these marine products are unrelated to both the fresh water supply and the land area, although the living entities in rivers, lakes and seas as well as ones on land are dependent on electron and H^+ from a water molecule.

Human beings are at the top of the food chain in the ecosystem and eat every creature as food, which this organic substance is basically formed with CO_2 and H_2O. According to FAO (1997), the quantities of fresh water necessary for producing 1kg of roots, grains, poultry meat and beef were 1, 1.5, 6 and 15 m^3, respectively. If we serve them by the hunting of wild fishery and animals, it is not necessary to meet a demand for fresh water, but it is necessary to have a suitably wide place for hunting and fishing. On the contrary, we need very large land area, if we are going to raise and let animals graze in a particular grassy plain so that there is little rain snowfall, and the productivity per unit area becomes low extremely.

In any case, humans must prevent the atmospheric density of CO_2 reaches a fatal level, for which the best method will be to contribute to the recycling society as biomass by covering the land by woody plants. If so, the first step of that purpose will be that humans distribute an area as wide as possible of the land on the earth for tree planting. For this, the stock-raising industry for grazing animals is unfavorable for occupying a very large grassy place, and the ranch will have to greatly reduce for the survival of humans. For this, however, there is some balance among essential amino acids, such as including a benzene ring in structure, which humans must intake from animal proteins.

For this, the stock-raising industry to occupy a grassy place only for the purpose of production of beef cattle on a large scale cannot but to decline. The breeding of cows should be limited to milk cows, and similarly, the sheep cannot but accept only wool production, too. At first, the state of the stock-raising industry in the future society must be reformed for the utilization of the land innovatively. The stock-raising industry to occupy a grassy place for only production of beef cattle on large scale will become extinct. Horse racing is destined to become the entertainment that will not be welcomed and will be limited to the support of human labor. After having finished a role, all of them will function as the animal protein for edible meat supply decreases.

In the future, however, the fisheries industry should be greatly concerned with supplying animal proteins necessary for survival, but it is uncertain at the moment. The reason is because I cannot expect how high the marine acidification and the seawater temperature will rise. Therefore, the concentrations of CO_2 and O_2 in the seawater decrease gradually in future, for the decreasing solubility of gases with increasing seawater temperature, and as a result the marine ecosystem will become poor. Probably, the sea level in the future seems to increase by the thermal expansion of sea itself and by the fusion of the permanently frozen ground, and to expand the sea area against the land area. However, I do not think the ocean which was acidified and warmed has rich marine products. At that time, human beings may live by getting the animal proteins from the culture type marine products, but not from the fishery type ones which will only be obtained when a limited sea area will be effectively used.

The space collection type production of animal proteins has already spread by the breeding of chickens, but such as the three-dimensional use of space, called the battery cage method, should be introduced not only for egg production but also for chicken meat production in the future. In the future, if human beings look for the higher efficiency of animal protein production, they should utilize the insects that are at the first stage of the food chain. We Japanese had made insects edible before the spread of pesticides. I liked the fried-up grasshoppers that lived in the rice fields, too. However, it seems that the weather prediction for the future will be increasingly higher temperatures and more humidity at many places so there are no farming places unrelated to a pesticide and no ranch with large grassland.

As means to secure the food for many people, a food production method by the three-dimensional land must be investigated while confronting global warming in the near future. We should think of the breeding methods of various kinds of insects, which are edible at the stage of larva or of imago. After having exhausted fossil fuels before long, we will not to be able to build a sustainable recycling society only by the superficial use of very large land. Humans must part from the ranch type stock-raising industry and the pelagic fishing or whaling.

By the way, an intake of the animal proteins is necessary for human growth, but it is a story on having been able to secure enough quantities of cereals. As a matter of course, that production depends on fresh water supply. According to the former FAO report, it is said that 1.5 m^3 of fresh water is necessary for cereal production of 1 kg cereals, being in quantity of 1/10 of the beef and equal to 1/4 of the chicken. Moreover, for growing food such as roots, vegetables or fruits, the amount of 1m^3 per kg of fresh water would be required.

In general, the higher plants constituting the base of a food chain in the natural world produce an ecosystem peculiar to the area, but the ways of their life are different depending on fresh water supply during their growth season. In addition, most plants have evolved adaptively in response to the length of the sunshine with a different endogenous clock according to the latitude where its kind was born. All plants avoid the dry season and change growth style when wet season comes. Here, it will be necessary to think by dividing the plants into two groups: one is the trees which function as a biomass by which tomorrow's sustained recycling society will be leaded, and the other is the foods of grains, roots, culinary plants which are indispensable for living. However, the woody plants of the former is also important as the source supplying steam into the atmosphere and supplying fresh water as rain or snow fall, and the plants of the latter are important not only as staple and side foods but also as the materials of medicine and textiles etc., and furthermore as baits for various kinds of stock-raising industry and marine cultured industry.

By the way, the biggest problem of human being is to control the CO_2 concentration below 1,000 ppm in the atmosphere in which is exhausted, in order to not disturb human oxygen breathing even if global warming advances. To that end, human beings must increase the forest cover as the CO_2 absorption source as much as possible. As the fossil fuels are worn out, the great plain of very large Siberia and the sterile ground in northern Canada may change into the place suitable for farming, but in order to hold the CO_2 concentration down to 1,000 ppm, in all land area, the earth must be blessed constantly with rain or snow falls. It must be assigned for forest creation more than for food production. What we should pay attention to is, what is the acceptable population number for the capacity on earth?

I have estimated that the population capacity of the earth would be 8 billion from those days of the first Earth Summit. According to UNEP, the world population is approximately 7.4 billion in 2014, may exceed my estimate soon, but the increasing rates of population begin to decline in many countries. Accordingly, the world population may converge in 8 billion or more. Moreover, there is no country that awfully suffers from starvation in particular at present. Although the self-sufficient rate of food in Japan is very low with about 40%, the values in China and India, both large Asian countries are at the top, also begin to decrease. In the case of Japan, approximately 40% of low situation have always continued, but the self-sufficient rate of food seems to begin increasing automatically to near 80%, since a decrease process of population has already begun with a low birth rate of about 0.35. These facts suggest that food will not cause any serious problem in Asia, if natural environments do not change significantly.

Nevertheless, it is extremely important that we prepare frequent occurrence of the abnormal weather with the global warming of the earth. And we must work hard to increase the forest now, although it is also important to suppress the discharge of CO_2, CH_4 and N_2O. At first, we must work hard at tree planting. In this case, there will be not only the choice of the tree species suitable for the place but also developing new varieties of woody plants. Secondly, there will be work to enlarge the land space for tree planting. In this case, there will be work to produce vacant land for planting by raising the productivity per unit area for farming. Finally, the most difficult thing is an effort to change the area that is not suitable for planting to land that is possible for planting. In this item, I want to describe about the trial of the second problem.

The food self-sufficiency rate in Japan has been coming in only around 40% for a long term, because the Japanese islands are in the steep mountains which are not suitable for farming. Near 76% of the country is covered exhaustively in the forest, which has already played the role to prevent global warming.
Japan became a small country after World War II, but the effort began since early times in order to raise the food self-sufficiency rate somehow. Our research group in which I participated wrestled for a study whether we could grow the plants in the completely artificial space that neither the soil nor the sunlight were used at all. We adopted rice and duckweed, both taxonomically belonging to the monocotyledoneae, as the representative plants which accomplish the original scenery of the farm village of Japan. It was a reckless challenge in those

days, but we ignored the air-conditioning and lighting costs, and received a bailout from the government as a fundamental research. We adopted two kinds of duckweeds which represented vegetables. *Lemna gibba* represented the long-day plant which has been adopted to the northern area, while *L. pausicostata* was representative of the short-day plants which was adopted to southern areas. From this research group, some extremely important research results lead to the birth and development of biology, agriculture and environmental sciences for the future.

The first findings by us were that *L. pausicostata* a short-day plant which could not form a floral bud only in autumn under the natural conditions could begin the reproductive growth as if it was grown in the culture solution in which the utility of Fe^{+} was disturbed and/or the concentration of sucrose was raised (1964, PCP.,5,513). Thereafter, the former finding became the discovery of mugineic acid (an amino acid making a complex with Fe^{+}) which controlled the absorption efficiency of iron ion in the soil, which was developed by Dr. S. Takagi of a cooperation researcher who was commended in the Japan Prize of Agricultural Science later (1988). His achievements opened up a course to promote the success of planting in areas consisting of alkali soil.

The second finding seems to be the origin of the development of today's plant nursery where vegetable plants were completely cultured artificially. Thereafter, we could succeed to separate two roles of the sunlight in the plants, in which experiments using two kinds of duckweeds were done under the condition containing sugar in the culture medium in order to exclude the action of the light for photosynthesis (1966,CPC.7,59). Regardless of a biological endogenous clock and with or without light, the duckweeds of both types could grow vegetatively and form the floral buds only changing a balance between red light and blue or infrared light under continuous illumination. The difference with the duckweed which adapted to the temperate zone from the duckweed which adapted to the temperate and subtropical zones was merely that the former was more dependent on the blue and infrared lights. Thus, we were able to acquire the theoretical condition that the higher plants, especially vegetables, were cultivated indoor artificially, if a lighting equipment having higher conversion efficiency to lights with different colors of electricity would be developed in the future.

These research results were utilized in Tsukuba International Science and Technology Exposition 1985 by Mr. Shigeo Nozawa who visited and studied at my laboratory. Here, he grew one tomato seedling like a tree (photo.1: Mr. Nozawa (the right side) and the author under the tomato plant which grew like a tree) and produced 13,000 fruits in half a year (photo.2,3). The trial of the tomato cultivation by him came out of the repayment of favor to the Middle Eastern people who have donated crude oil to Japan.

The origin of his idea allowed people who lived in drying places to eat fresh leafy greens anytime. In order to grow vegetables with a quantity of limited fresh water, the plants were grown in the nutrient solution containing minerals which was bubbled by air to supply O_2 and CO_2^- to the root system.

Fortunately, Japanese scientists of the Wining Nobel Prize two years ago succeeded to convert even a blue light effectively. In Japan, thus, the three dimensions cultivation of culinary plants began to spread like the breeding of chickens. This situation means that we can distribute land area into the forest for absorption retention of CO_2 as well as raise the food self-sufficiency rate of Japan, so long as Japan can maintain a stable energy power supply in the future of the fossil fuels drying up.

1
2
3

7. The Human Duty through a Period before the Drying up of Fossil Fuels

According to the 4th report of IPCC in 2007, the sea level rises in the speed of 3.1 mm per year for 10 years from 1993 to 2003. Also many countries report similar rise of the sea level, though its levels are different, which has been thought to be caused by the fusion of the ice field and seat at the polar regions and the higher mountain and by the thermal expansion of the ocean itself. The erosion in the coastal place advances with global warming, resulting in gradual reduction of land area, which suggests that the relative decrease of the land has become serious. The countries which have already been on the verge of submergence cannot but be buried in sea water in the future.

As the marine area spread, it brings more and more atmospheric steam pressure so that it will increase the frequency of abnormal weather. The ocean is also one of the absorption sources of CO_2, but we cannot expect the great role of the ocean as the temperature of seawater increases and the CO_2 solubility decreases. Besides, since CO_2 acidifies the ocean, the ocean born in this way does not seem to be a favorable environment for human beings in the future. Probably, the acidified ocean would destroy existing marine ecosystems, and result in new ecosystems that will not be preferable for human beings. It will only take away land space from humans for tree planting.

Until now, I have written about the absorption sources of CO_2 that is the main constituent of the global warming gas. By the way, I cannot but overlook it once about the discharge source, since CO_2 concentration in the atmosphere is different in quantities between discharge and absorption. As mentioned above, China which has represented the developing countries came to already be the world's worst CO_2 emission country, 26.9% of 31,800 million tons of total discharges across the United States (EDMC, 2014). They say the reason was that China was not equal to the developed countries with the CO_2 discharge per one nation, but the reality that it had surpassed 1/4 in the gross weight would let China feel that it was impossible to still continue a conventional attitude. I would be happy if there was a chance to compromise in the developed countries.

If more than 1,300 million Chinese are going to aim at life improvement to a level appropriate to increasing GDP, the breaker of the global environment couldn't but criticize the Chinese. In the times of Kyoto Protocol of 1997, the CO_2 emission from China was in second place of the world next to the United States. As China's GDP increased, however, China became the discharge country in first place of the world. At that time of 2000, China poured power into economic growth, at which the CO_2 emission was big in the industrial section, 38% from the electric power generation industry, 22% from the steel industry and 11% from the cement industry, respectively (WRI. 2005). Interestingly here, there is the cement industry in the first step which separates CaO or MgO by heating for decarboxylation of $CaCO_3$ or $MgCO_3$, $CaCO_3 \rightarrow CaO + CO_2$ or $MgCO_3 \rightarrow MgO + CO_2$, thus accompanying of CO_2 discharge. Thus, the cement industry causes the discharge of CO_2 with large quantities. The fact that China, unlike Japan, highly depends on the cement industry, it seems to be proof that China had concentrated power on infrastructure maintenance. In Japan after 3.11 great earthquake, the fossil fuel dependence on generation increased by the nuclear plant accident, the CO_2 discharge from the generation section increased and reached 39% of 127,600 tons, but it is 26% from industry in which 13.6% was from the steel industry and surpassed one from the cement industry (JCCCA,2014).

By the way, cement is a material for house construction that humans cannot miss for survival, and it is the material of the infrastructure maintenance which reconstruction of the earth predicted in the future. As well as fossil fuels, both limestone of $CaCO_3$ and magnesite of $MgCO_3$ which were raw materials of the cement industry were formed from CO_2 which was an atmospheric major component in the days of the earth. In addition, most of CO_2 which existed changing a figure into this inorganic matter rather than fossil fuels more in those days, and they were supposed to build land. Both materials made from ancient CO_2 have contributed to humans as ones being indispensable for survival. The first living entity was born in a process during which the concentration of atmospheric CO_2 decreased and the concentration of atmospheric O_2 increased. Humans had finally evolved when those concentrations in the atmosphere became most suitable for its survival. Accordingly, the survival of all living entities which accompanied evolution is not permitted if even CO_2 begins to return to ancient level, since gas exchange on respiration is disturbed.

For example, even IPCC (AR5, 2014) sets 4 kinds of CO_2 concentration of 450, 550, 650 and 750 ppm and predicts the changes of various kinds of factors in each case including cost. However, I don't think that it is much about CO_2 substitution reaction in breathing oxygen here. Besides, I don't think a significant discussion without a negative thought resulting from 3・11 disaster in Japan, which is resistant to the forthcoming atomic energy use afterwards. Their argument was based on assuming the world population of 9.5 billion people, but judging from the spread of education improvement of life, I still estimate 8 billion lay figure to remain the duty of humans living in the present age that we should not let earth's CO_2 density exceed 1,000 ppm in the future either.

While existing fossil fuels are useful, it seems that the atmospheric CO_2 concentration will continue increasing. When a situation of the drying up begins to be visible, however, the prices of the fossil fuels will rise, and their role finishes as the fuels and must carry a position rather as a precious organic production material. If it is that time, the role as the portable energy of fossil fuels must be replaced by hydrogen gas and electricity. The times when all cars loaded with a H_2 gas tank or a charged battery in place of a gasoline tank must be approached and run near at hand. That is, the times when all depends on the fossil fuels for a driving force will be over in shipping industry and in agriculture soon. And, also the fuel of the H_2 gas will be fresh water again.

In addition, the reality that the atmospheric CO_2 concentration continues climbing for a while gives the possibility of the rapid increase of the photosynthesis efficiency in plants. Therefore, the production of the agricultural efficiency may improve drastically, if nutrients in quantity theoretically appropriate to the atmospheric CO_2 level are supplied and if the sun light in quantity is enough to reduce CO_2 into irradiated organic matter. If global warming is followed by rising temperatures during night, however, the consumption of produced organic matter in the night must increase. Accordingly, the developing new varieties of crop plants with new high temperature tolerance, and further with disease tolerance, may be necessary.

In any case, CO_2 was in excess in the atmosphere, and if there were all essential elements necessary for plant growth in the soil, it was revealed that the determinative of the population capacity of the earth was not food supply. It was thus revealed that the determinative was limited to light of the sun and a fresh water supply at population capacity of the earth. As for the element which brought the extinction of humans on earth, it may be said that both food shortage and the radioactive contamination are unrelated. If I tell others, the direct factor causing human deaths may be abnormal CO2 accumulation in the air which humans themselves scatter as waste. When the CO_2 concentration reaches 1,000 ppm or higher, humans will trace their way to extinction.

I don't know whether the peoples of the developing countries accepted talks with the developed country side in COP etc., after understanding the facts that the CO_2 which was exhausted by human activities and accumulating in the atmosphere brings not only global warming but also the fatal risk that may directly cause death. In any case, however, I am thankful for the people of developing countries having noticed that it is not a situation to keep forcing responsibility on the developed country side. An international organization as represented by IPCC must let the talks of the framework of CO_2 emission right postpone a period before atmospheric CO_2 concentration in air reaches a fatal level. Meanwhile, this is what humans should do for the next generation.

8. From the Nicaragua Canal Plan to the Construction of a Canal for the Aral Sea

At the first Earth Summit in 1992, the main theme was about how humans could continue living forever on the earth with limited resources, if we did it. The social state investigated there was a sustainable society by the circulation use of resources. Here, I have stated that the basic required factors are the energy of the sun and the supply of fresh water (H_2O), because both CO_2 and O_2 molecules, which are indispensable for survival, have been continuing to exist in the atmosphere. However, if each gas exists extra, it is harmful together. The increasing concentration of CO_2 in the air brings not only global warming, but it is also fatal if its concentration rises by 10 times of its current concentration. On the other hand, the O_2 is accompanied by peroxidative reactions in proportion to a surge of the concentration, and its accumulation which is dissociated from existing water and hydrocarbons etc., promotes the deterioration of all materials and shortens life. I suppose that the brilliant sun exists semi-permanently. So, the final problems are narrowed down to what should be done for humans to maintain the current atmosphere composition by sharing fresh water resources.

Of course, the various kinds of international organizations led by the United Nations have repeated an effort, and at last, seem to be reaching an agreement in a common direction, including developed and developing countries, except some selfish people. However, I have developed an argument for myself here, because I did not think that their efforts were enough without a figure of the earth after having used all of fossil fuels exhaustively. And it is the action, which should be done during a period while the fuel fossils are cheap and useful as energy sources as well as resources of organic matter.

By the way, only the CO_2 release derived from limestone as for the cement production seems to continue as far as humans live, since the cement is an essential material for e humans, even if fuel fossils dry up. That is, it is thought that even if the fossil fuels dry up the upward trend of the atmospheric CO_2 concentration is maintained thereafter, and that the high recycling of element C may not be necessary for the production of biomass and organic matter in the future either. If so, I think humans should make power while never raising the atmospheric concentration of CO_2 above 1,000 ppm, but not about the recycling use of the C source. Then, I want to imagine the social structure of the coming times.

At that time, how to get an output of energy and how the supply of fresh water would turn out were important. Energy does not have meaning if it is not always supplied steadily. Both fossil fuels and uranium resources will disappear by consumption together, however, it is natural that it will become the times when humans cannot but entrust all of the important primary energy sources to natural energy. The sources of natural energy are as follows: sunlight, the flow of wind, the wave force of the sea, tidal power, the flow of river, a tide, territorial heat, and the difference of temperature and the difference of position etc. Most of them are not a stable output source and are influenced by the sun. The probable stable energy sources will be limited to terrestrial heat, the flow of river and the energy of the position.

By the way, is it good to depend on geothermal heat as a stable energy output? With Japan's volcanic islands, there are many experts recommending geothermal heat dependence of the power generation in Japan. In the school of Japan, however, the eruption of the volcano as well as the giant earthquake have been taught to occur when the plates constituting the earth crust collide each other by moving with a mantle convection and cause friction, in which the mantle convection is caused by increase of the density by the cooling in the surface of the earth's neighboring areas. Therefore, I think that excessive dependence on geothermal heat of the power generation may increase risks of natural disaster and should be avoided even if it is a stable output.

Conclusively, future humans cannot but depend on sunlight to be influenced by weather with day and night for most of the primary energy output. The important problems here are how the unstable energy of the sun can convert to stable second energy output, and further how it can be done cheaply.

Besides geothermal heat, there is nature which can output the primary electricity steadily, which is the wave power existing through the year regardless of day and night. However, the estimated generation gross amount is not large, and then it may carry on only as a role to maintain the life in the remote islands.

Another marine source, there is tidal power occurring from moving water between low and high tides, but the output is not the stable power supply which humans can use promptly since it is influenced by a periodical movement of the moon. Moreover, the marine energy does not seem to be for a general purpose since it is limited in a country only blessed with seas and it is planned under coexistence with the coastal fishery and with the security of the shipping route.

Further, the quantity of total generation though to be stable is small for a general-purpose supply, and so it is thought that the countries that can use it are limited. In the ocean, the generation (Ocean Thermal Energy Conversion) using the temperature difference between the outer layer and the bottom layer in the sea is only the generation using sunlight indirectly, but it is attractive as the solar energy absorbed in the ocean is saved even if it is at night. But, global warming may raise the seawater temperature in the depths of the sea area before long. In that case, the temperature differences with the depths may decrease too, resultantly following the electromotive force declining.

This possibility suggests that the ocean having higher specific heat capacity is also inappropriate for the retention of the primary power supplied in large quantities from the sun. Then, the work that humans should wrestle on now will be how we could convert the unstable primary energy from the sun into stable output at the secondary energy stage, and how we could decrease the cost as low as possible. Humans must achieve those observances while the prices of fossil fuels are relatively cheap.

(A) Conversion and retention to stable energy from unstable primary natural energy:

It seems that the atmosphere of the earth which performed warming includes H_2O as steam at a large quantity similar to CO_2, CH_4 and N_2O. Accordingly, it may be local, but the atmosphere circulation becomes intense, and the total quantity of rain and snowfall through a year will significantly increase at the times when most fossil fuels will dry up. Basically, this rainwater and melting snow water should be stored in the dam for pumped-storage hydroelectricity (PSH), excepting a part to not annihilate the ecosystem peculiar to the area. The fresh water saved in the dam is natural to be used for waterworks, agriculture and industry, while the remaining can be kept mainly to store the primary energy of the sun alive as a stable second energy source repeatedly. The PSH which has already spread all over the world can store energy from photovoltaic and wind-generated electricity origins in the form of gravitational potential energy. The extra energy from the natural electricity can enable to steadily supply energy throughout the day.

In this case, we must maintain the biotope for increasing types of fish without making much of a commensal principal with nature, for which the setting of the dam should wait in some branch of a river. In addition, the establishment of the movable outlet of the sedimentation soil and sand into the PHS dam will become necessary for the extension of its life, and the prevention of deterioration of the concrete structure body in itself will be necessary, too. With that in mind, PHS must be built assuming the harmony with natural environments positively in the future. Besides, on the update of the dam where life came to, the PHS should be adopted without words and the dam lake there is good. It will be used for self-sufficient food supply by the culture of freshwater fish.

(B) Conversion and retention to stable energy of H_2 gas from Unstable primary natural energy

The hydrogen (H_2) gas which is already called zero-emission fuel is used as a power source. And a study begins on how to use it for electricity generation recently. Now, all car manufacturers push forward practical use in electricity or in H_2, as a power source of the next-generation car. The electric car in the former will carry many large-capacity batteries and can make use as a temporary bank of electricity transmitted from the primary generation station in each family. Probably, the car with the batteries will be more desirable in the countries which cannot build the PHS mentioned above. The life span of established batteries is short unfortunately. Therefore, it is desirable that the variable and intermittent electricity depending upon various natural energies should be stored in each family and should be used after being gathered as a society, thus setting a supplying net for sharing electricity altogether.

However, the fuel cell-powered car to hold a H_2 tank may be suitable for practical use in Japan and each Southeastern Asian Country where many PHSs will be able to be built. Unfortunately, most of H_2 gas is produced from natural gas as raw materials now, merely decreasing the discharge of CO_2. We have already greeted the times when we must fix the system which can produce H_2 gas by electrolysis of the fresh water using natural energy as soon as possible.

In countries lacking fresh water resources, development of a system to separate fresh water from sea water without energy load must be hurried. The best method will be to introduce seawater into the slight aquarium in the closedown glass room, to concentrate by solar heat in sequence, and to isolate fresh water and inorganic matter included in the seawater, in which evaporated steam will be condensed on a glass surface in the night time and be stored in a basement tank as fresh water. On the other hand, the sediment deposited at the bottom of the evaporation pond is collected, transferred to another room and provided as inorganic raw materials because it contains various inorganic elements. Most will be NaCl, but it will also contain various industrial raw materials, which their costs will become very cheap because their production is performed only with the solar energy in the glass house of which the deterioration speed is extremely slow. In the land suffering from salt breeze damage, this kind of glass house may be utilized for drinking water, not to take out H_2 gas. I want to describe about the residual handling included in the seawater again in item 10(A) later.

The current city is constructed by a premise to set various kinds of service lines on the road. In the present, the natural gas, mostly CH_4, is supplied to each family through the underground plumbing mainly from each local gas center. However, the risk of hydrogen explosion is very high, and even if a fire explosion occurred, we cannot but keep it locally. Even before the drying up of fossil fuels, in the countries having a lot of natural energy, it is thought that H_2 can be used as a raw material which is an important reducing agent in industry, which is rather indispensable for the first reaction step on organic matter production.

(C) The evolution on planet earth was adaption to O_2 concentration. How can humans fight against extra O_2 ?

The earth was born as a member of the solar system, but the O_2 concentration included in the atmosphere was only about 2 ppm to occur by the dissociation by ultraviolet rays of the water for a while. 2,700 million years ago, the first living entity which could photosynthesize started on earth, breaking down H_2O to H^+ and O_2, in which the H^+ was used for the production of organic matter through the reduction of CO_2 included in the atmosphere in large quantities, and the O_2 was released in the atmosphere. Afterwards, the evolution of the creatures began as adaptation to a surge of the atmospheric O_2 and produced variety, and the humans would be located at the top. During this period, the atmospheric O_2 concentration increased to approximately 10^4 times, which became steady as the strong atmosphere of oxidation power. As creatures prospered in this oxidative environment, plants evolved angiosperms and animals evolved mammals, in which they grew without touching the extra O_2 and were able to carry on in future generations.

Judging from a different viewpoint, the act that humans burnt fossil fuels and resultantly reduced O_2 concentration might be significant at the point that controlled a menace from O_2 poisoning. However, fossil fuels will disappear before long, and resultantly the atmospheric daily O_2 consumption will seem to decline with the change of people's lifestyles, shifting from fossil fuel dependency towards a dependency on electricity and H_2 gas. The H_2 dependence society may think with an ideal form of complete resource circulation, because the H_2 gas which was from H_2O discharges only H_2O on the occasion of combustion. In the hydrogen society, however, the H_2 was used also as a reducing agent for the production of various kinds of artificial organic matter, in which only O_2 was left unused.

On the other hand, woody plants release O_2 more than the absorption quantity of CO_2 in a photosynthesis process, for which the quantity of the released O_2 increases as forest cover is expanded. And it was a cause of increasing the atmospheric O_2 concentration after the birth of earth, and was a background of the evolution to bring biodiversity. The increase of an atmospheric abnormal level of O_2 is undesirable, but humans must work hard at planting to prevent global warming in order to promote absorption in CO_2. Therefore, I think that the future atmosphere of the earth may become an oxidative space in enormous O_2, which is not a favorable environment for human beings. As for the reason, the deterioration of the materials with time as well as the aging of creatures depends on an oxidative reaction basically. Accordingly, it is important that humans should challenge while we can, using fossil fuels to suppress the rise in CO_2 level 1,000 ppm without increasing the present O_2 level.

Unfortunately, I do not have the wisdom to retain O_2 safely. The organic matter which had extra O_2 in a molecule is generally fragile and may not retain O_2 molecule as organic matter. The possibility cannot but depend on research and development in this field. In the case of inorganic matter, it is basically the same. The inorganic substances are unlikely to be stored in the peroxidative state. Besides, the earth itself includes O_2 in large quantities, 70% more, which suggests that there may be no element which can combine with O_2. Aft H_2 by some kind of method, we always must think about the er all, when we take out means to leave it as H_2O.

(D) No reduction and retention of CO_2 from the Atmosphere Without consumption of energy.

Vast energy is necessary for collection and retention of main global warming gas CO_2, nevertheless humans must work on its collection and retention using the energy from fossil fuels to hold down global warming. The detailed report is reported in IPCC 2005. This report introduces various kinds of CO_2 collection and retention methods to restrain global warming action of CO_2 exclusively in detail. However, because an important point in this report focuses on the global warming aspect, different thoughts come out when another important point focuses on the relation to health of human beings in CO_2 itself.

As reported in LIMA COP20 two years ago, the prediction

that the atmospheric temperature would rise about 2℃ by 2050 if we do not do anything. It is very serious. Probably the island countries do not only become extinct by a rise in sea level, but also the coastal erosion could threaten the life of people in each country. If so, I can understand the circumstances that humans must regulate not only the CO_2 discharge, but also must collect and store the exhausted CO_2. However, I do not think that the technique adopted now in oil wells and gas fields is good to apply for a long-term retention of discharged CO_2. There, the collected CO_2 is liquefied by compression and then sent underground under pressure. The long-term retention of CO_2 is said only for the sedimentation tray of the stratum which is deeper than 800 m below ground. As well as the disaster by the great sprout of CO_2 in Africa that was mentioned above, the leak of natural CO_2 has been observed in many places in the world. Judging from these facts, I cannot evaluate that the retention method with the risk of CO_2 leak, even if the retention site of CO_2 is land or at the bottom of the sea. This may affect human lives by changing the future atmospheric composition after several thousands of years.

Until fossil fuels are being used exhaustively, humans will make an effort to hold down an atmospheric CO_2 concentration to a level not to exceed 1,000 ppm for health maintenance. However, I cannot say at which level of CO_2 is able to keep its steady state after fossil fuels have been depleted. There is risk for leak by natural disaster explosively only at one sweep if humans shall be in a condition to have to spend under the higher concentrations of CO_2 that are almost 1,000ppm, even if the retention site of CO_2 is deeper than 800 m. If humans live in an atmosphere at high concentration of CO_2 near 1,000 ppm, the dying scenario is possible without being able to respond to a large-scale of CO_2 leak in a wide range of areas. If a retention site is the earth layer at the bottom of the sea, the CO_2 leak may invite the acidification of the deep ocean sea, which may have a fatal influence on the marine ecosystem.

In addition in IPCC2005, another wisdom to make marine space as a retention site of CO_2 is introduced. After multiplying by means of fixing atmospheric CO_2 by photosynthesis of the ocean climate phytoplankton, the multiplied plankton cells are collected, then solidified after drying, and then are sunk as a big solid body in the deep sea. Also in this manner, human beings are described to be able to live by isolating atmospheric CO_2 of higher levels for a long term. By the way, the problem of the first Earth Summit in 1992 was to investigate a state of the sustainable future of humans and one of the resources which needed recycling was phosphorous (Pi) of the nutrient which was indispensable to production of foods. The phytoplankton sunk in the bottom of the sea means that Pi- elements of vast quantity was excluded from the desirable recycling system. Humans cannot agree to the recycling use retention method of the CO_2-reducing quantity of possible Pi, because Pi which the plankton held is isolated in the bottom of sea.

Anyways, there is a C element like O element in large quantities on the earth from those days of earth's birth, but the extra existence commits fatal action together and the retention of C must be formed in the future environment so that humans can live peacefully and safely. At such a point of view, the retention of the element C jumps out in large quantities accidentally as CO_2 is basically undesirable. However, it will be to say that CO_2 retention, for the society to be able to take out as raw materials from organic matter products as needed is desirable while holding down CO_2 in the atmosphere to less than 1,000 ppm. To that end, the method that we can evaluate will become the CO_2 retention that we transfer CO_2 in some kind of stable form during a long term, and it is made in the interval where humans still hold fossil fuels as mass energy. In addition, the very large space where it is convenient for humans is also necessary. I want to describe it in Item9. Remaining one is a biological method to shut CO_2 as coral in the ocean and as wood in the land. As these are fixation and storage of CO_2 without consuming artificial energy, I think it is an extremely important action not to miss in the future of earth.

(E) Long-term retention of CO_2 as inorganic carbonates and stable organic matter

As atmospheric CO_2 nears 1,000 ppm, CO_2 is becoming a toxic substance which may disturb CO_2 substitution in the aerobic respiration of animals and plants. Accordingly, even if humans continue to substitute for a natural energy in all fossil fuels, and even if it stores in the underground and at the bottom of the sea with CO_2 in molecular structure, there is hardly any meaning because of no decrease in the carbon element that the earth can hold. Therefore, even if humans retain CO_2 in the bottom of the sea and in the deep stratum keeping its molecular structure intact, humans cannot avoid the risk by the great sprout of CO_2 in emergency. Humans must not seem to live in fear of atmospheric CO_2 with a poisonous gas. For this, humans must convert the CO_2 molecule into other materials while it can use fossil fuels as a heat source. The prospective material does not have the toxicity, and can be kept stably for a long term, and further itself as a useful material or a raw source for production of other useful materials. Or, it may be the material that is possibly useful to humans in the future. Anyway, humans must let CO_2 change the figure.

The materials fulfilling the above-mentioned conditions with inorganic matter are magnesium carbonate ($MgCO_3$) and calcium carbonate ($CaCO_3$). Those raw materials are included in the ocean as silicates, and also as the components of rocks, for example, an amphibole mineral consisting of calcium magnesium silicate. The representative mineral was asbestos in so-called silicate minerals which had been used for building materials as a toxic substance all over the world until it was prohibited. As asbestos causes the inducement of health damage, it is extremely serious. It would be good if we could use the asbestos as raw materials for retention of CO_2.

At first in it, we must build the exclusive factory producing carbonate salts, even if the raw material for this is harmful asbestos or silicates melting in the sea, while letting it cooperate with a large-scale retention site even if it is useful partially. Moreover, vast energy is indispensable for the preprocessing of raw materials, for pyrogenic reactions and for waste port. Besides, if we assume a retention site in the bottom of the sea for a long-term storage, both $CaCO_3$ and $MgCO_3$ must be formed on the cube after solidification for transportation, and the surface processing will be necessary for storing at the bottom of the sea, a long-term retention.

On the other hand, the CO_2 fixation by organic chemistry is artificial photosynthesis, in which there should be some kinds of materials having higher functions than the above inorganic CO_2 salts. As it is light of the sun-dependent and finishes all energy, it must occupy the important situation that takes a part of the resources recycling society after the fossil fuels dry up. We are in the times when we can acquire indispensable H^+ from electrolysis of the water using solar energy easily to reduce CO_2 at the time, and to fix it in organic matter. Or, because the water evolved H_2 gas is freely obtained, we can produce stable organic materials suitable to CO_2 retention by adding CO_2 molecules to some biomass, or useful organic materials similarly.

The chemical fixation possibilities for CO_2 are similarly applicable to CH_4 that also takes more part in global warming by absorbing the solar lights in infrared wave zones in large quantities than CO_2 does. Therefore, the biggest problem is how we can cheaply capture both CO_2 and CH_4 included very slightly in the atmosphere, and how we can guide them to the levels at which the chemical reactions proceed smoothly. The separation of both gases has been already carried out in the natural gas field, but it is extremely difficult to take a CH_4 gas being contained in ppm order of one column level from the air effectively, whereas

CO_2 is only three columns of ppm levels. Of course, it is allowed to take both out from all over the air without isolating into each gas, and then may separate and concentrate each. In this case, we may just apply means used in the gas field now.

Thus, three ways of means are thought about the capture and storage of CO_2 in the air. Firstly, the fixation to carbohydrates of CO_2 is through the artificial photosynthesis, but I do not think storage of semi-permanent CO_2 is suitable, though being usable for recycling of the alcoholic fuel. The second way is going to store CO_2 by tying it to some partner additionally, in which the organic matter produced become different by a kind of partners in biomass. The third way polymerizes CO_2 by reducing it to CH_4, thus producing some kinds of hydrocarbon-based materials. However, the reaction of any case does not begin in ppm order in the air.

Therefore, I think that humans should make power for the third way arresting not only the capture of CO_2 at the same time but also the capture of CH_4 which acts to cause more efficiently global warming than CO_2 does. All atmospheric composition ingredients on the earth are fixed at temperatures between the ocean and the land by equilibrium of absorption and release to change with temperature of seawater having a higher thermal capacity. The decrease in solubility to sea water of CO_2 has already begun as chlorosis of the coral as a drop of CO_3^- ion content in seawater with warming above all. Humans must hurry collection of atmospheric CO_2 gas to evade a serious blow to the marine ecosystem. But I cannot agree with the capture of both CO_2 and CH_4 using the membranes which must be often changed with expensive new ones. The reason is that the membranes are quality of organic compound prepared with fossil fuels as resources, and the sustainable environment managed with only natural energy in the future must be controlled with materials which can reproduce at low-cost and natural as possible.

The concentration plan of CO_2 that I have thought about may have been already put to practical use. If the separation and concentration system of an atmospheric gas component as follows is completed, it should be operated as soon as possible in each place around the world. However, I have not received any news that the factory which I have thought about was built. At first, the air purification factory needs to be built with UNs leadership in the coastal place of the world. Further, the global community should fix the system to cover a bailout, to run it with a frame of a global community.

On the occasion of the construction of the exclusive factory building, the intake of the air is attached in every direction of the building so as to introduce easily the air depending on the direction of the wind. Needless to say, the fan turns if windless, and the inflow is reduced if winds are strong. At first, soil and sand grains are removed from the air through a filter layer. Thereafter, in order to trap nitrous oxide (N_2O) gas included at about 310 ppb in the air, the air is bubbled in the cistern containing diluted sulfate solution. Next, the air including CO_2, SO_3 and CH_4 are bubbled into a liquid phase of the caustic soda or potash over two times. In this step, CO_2 and SO_3 gases are trapped as the salt of Na or K, being ionized to CO_3^- and SO_3^{2-}, respectively. Before treating those at the next stage, the air including the remaining gases such as hydrophobic CH_4, and Freon, are carried to the dehumidification room for cooling dehydration, and then it is further passed within long passages packed with Silica Gel grains or calcium chloride crystals for removing more water. After exposed to the complete dehydration, the air including only the minor global warming gases is drained through a passage combined tandemly with the pipes which are filled with an active carbon and zeolite.

Under such a flow system, each targeted gas ingredient may be separated from the air and collected. The N_2O gas is capable of collecting from the sulfate solution, and the CO_2 gas is capable of collecting from the alkaline solution by acidifying it, although the separation from SO_3 gas is necessary. The technique of the gypsum ($CaSO_4 \cdot 2H_2O$) production is applicable to separation with CO_2 gas and SO_3 gas, where the mixed gases are sprayed with a limestone ($CaCO_3$) slurry in the middle of the duct and then are exposed to O_2 of the required amount, as a result producing a concentrated wet CO_2 gas. Thus provided pure CO_2 may become the raw materials of the organic matter production.

On the other hand, the active carbon and zeolite are moved into the case for heating linked with a pipe to a gas tank before taking out CH_4 gas and Freon gas. After the atmosphere in the case is substituted by the inert N_2 gas, the case is subjected to slight decompression and then heated moderately to leave CH_4

and Freon gases from the active carbon and zeolite. Here, we can catch CH_4 and Freon at the same time, although the concentration of the latter in the air seems to decline because of the prohibition of international law. Then, using the great difference in the phase transition point of Freon gas and CH_4, the CH_4 gas is able to be separated from Freon gas by letting gas go through in the pipes cooled to about -60 ℃ where the Freon is liquefied. Thus, not only CO_2 gas but also CH_4 gas which is the raw materials for organic substance production *can be collected in* each tank through a common flow system. If CH_4 is subjected to O_2 released on the occasion of electrolysis of water under a heating with a catalyst, CH_4 becomes methanol and will contribute to carbohydrate production. If CH_4 gases are heated under a high pressure, CH_4 itself becomes ethane which polymerize each other and will contribute to hydrocarbon production.

Probably, it will be said that the reproduction flow system of the atmosphere that I have suggested here is common knowledge. In addition, many companies are concerned with the reaction process wherever they may be as mentioned above, and they may finish those technological developments. But it is also obvious that if some profit is not promised, any business agreement is unlikely even if each step of the flow system such as above mentioned is synthesized. The important thing is that earth itself may perish. Humans must try to share thoughts and cooperate with each other for the reproduction of the future earth, cross the differences of creed and faith at the current time when the price of fossil fuel is relatively low. For this, the UN should take the leadership.

Because the alkali waste fluid adopted at the second stage of this flow system is neutralized with HCl, anybody should not pollute nature by discarding it into the ocean. The used active carbon, zeolite and silica gel are able to be used repeatedly. Moreover, humans can produce necessary organic chemical matter even after fossil fuels dry up. In addition, it seems that the need of the waste retention site of CO_2 and CH_4 will fall if we will incorporate CO_2 in some kinds of organic polymers which are hard to deteriorate. If those polymers are used for clothes or furniture, the reuse of atmospheric CO_2 is all right once, which is a good story if a consumption rate of the fossil fuels delays.

Nowadays, it works for maintenance and upbringing of the forest. However, the forest is changed into farmland and is cleared for the production of paper and building materials so that it continues decreasing the area. If the atmospheric CO_2 is able to make use in production of an artificial wood as well as a carbon fiber, it would be a chemistry technology dream to be able to keep it alive as a retention site of CO_2. The candidate of the materials may be a compound which is similar to synthetic graphite. In fact, it is an extremely useful industry material, because it is kept alive by an electrode of the lithium battery or car brake material. However, collection for the graphite of CO_2 may not be an unrealistic choice, because extreme mass energy is necessary for its composition. It is extremely lucky that the advanced nano-cellulose industry was recently born in Japan and it began to spread world-wide recently. Not only is nano-cellulose a material having a characteristic superior to the carbon fiber, but also its production is not influenced by the drying up of fossil fuels because the product from the industry uses plant fiber itself directly. However, the raw materials of this industry are only plant fibers, and so it is not established in the area that cannot secure the fresh water supply. The best raw materials for nano-cellulose industry are wood and bamboo, and those plants continue to take CO_2 from the atmosphere, and release H_2O to the atmosphere semi-permanently. At that time, a convention of the steam will be born in the continent inside, and ancient nature may revive.

(F) Climate remodeling of central Asia by the regeneration of the Aral Sea for renovation of forest.

The earth is a blue planet from satellite photo, on which the central part of the very large Eurasian Continent and the part of North Africa distinct are tinged with reddish color. If we photographed it long ago with both areas, however, it seems that the whole earth would be colored with a blue photograph which is imaged from the existing remains of an ancient structure which the human left for those areas. This is because humans have destroyed a conviction in the atmosphere of the water derived from the transpiration of forest on the construction of the above-mentioned the Great Wall of China or of grassy plain under the hyper pasturage. Since desiccation in the Eurasia Continent where the most serious one occurred was a man-made disaster, it may be said that humans have a duty to make an effort in order to leave it for descendants after working hard to regenerate the drying land on the green earth, while humans can still use fossil fuels as energy.

What humans should do now is to revive forest areas to absorb atmospheric CO_2 and to store trees. Thereby, the troposphere of the water will be born again in the continent inside, resulting to call back rain and snowfall. The most important factor indispensable to regeneration of the forest is moderate rain and snowfall. The only condition to make a definite promise of it is the existence of a lake surface and/or an ocean surface that exist in the continent inside at very large areas. To accept it, humans have responsibility for only the regeneration of the Aral Sea which is about to disappear, accomplishing with irrigation for farming in the 20th century.

Humans now have the technique to change the land of the drying into abundant land if blessed with fresh water also, and to prepare both foods and the forest, although influenced by soil constitution and ingredients. As shown in the photograph previously, the tomato plant can be grown like a tree if it is cultured in the nutrient solution with enough O_2 and CO_2. In other words, if there is just fresh water, the higher plants absorb atmospheric CO_2. And if it is a tree, it can continue accumulating and fix CO_2. Then, this is how humans can create such conditions.

9. Mr. Nozawa's Dream Living on the Tomorrow's Earth

Through a past statement, I have arranged that the key to hold the future of humans is the light of the sun and the fresh water, which is natural because the earth is a planet blessed with the water of the solar system. And on this earth, it was the forest to have been born during a long period of 2,700 million years while being with light and water. The forest have changed its own constitution according to the O_2 concentration that continued increasing in the process, and brought about various ecosystems and finally brought up the present ecosystem to have the human in the top. All of the experts agreed that tomorrow's leader of the global environment is the forest.

According to IPCC 2007, however, there are few areas where the land can be covered with trees. As presented here, unfortunately, it was humans to have had the earth consisting of dry plateaus. If so, what we should do is to regenerate earth itself while the price of the fossil fells is still cheap and useful. The original scenery of the earth which we should aim at will be a figure of the earth approximately 200,000 years ago, when the present human, *Homo sapiens*, was born. We will turn a terrestrial globe for such intension. There are 5 drying continents colored with yellowish brown. Four continents of the earth have very large dried areas dyed in yellowish-brown except South America. The Australia continent seems to be a place of drying since birth, since the ecosystem consists mainly of the plants and animals being strong in drying conditions.

(A) Betting the regeneration of desert at an irreversible stage on Mr. Nozawa's dream

By the way, most area of the north side in the African Continent which has been considered to be the ground of human birth had become the ground of the dilapidation represented in the Sahara, and it has been thought that it reached an irreversible stage. The area has passed the stage at which humans could regenerate by giving a hand to the original nature. However, there remains a slight possibility, which is to introduce Mr. Nozawa's dream shown in the photograph earlier. That is, his dream can come true by laying pipe lines to carry seawater from the Atlantic and/or the Mediterranean Sea to the central part of the Sahara, but it should be start it while humans can use the fossil fuels.

But there are some conditions to realize his dream concretely. Firstly, people to approve of his dream and to call together must be completely freed from faith at the very beginning. The freedom of religion should be guaranteed as human rights, but must be people who can keep everything about faith at home. Men and women must turn up, and to be able to walk towards a common aim together spreading it once if they stepped forward to one step of the house. Secondly, people living there need to have a clear vision about the residual disposal method of the seawater which will continue depositing with time. The suitable useful resources are included there, but excess NaCl continues depositing. Thirdly, people living there are able to accept the population capacity which faces margin with each other according to the food self-sufficient ratio. The people must build up society depending on the expansion of the scale and on the increase of the operation site. The reason is because it is really unknown how the expansion of the system of Mr. Nozawa's dream influences a climate of the North Africa. I would be happy if his dream would succeed to cause a climate change in the area.

To make use of his dream in earth regeneration, humans must not be particular about the style of a plant factory spreading in Japan now. At first, the seawater which has been carried to the central area of the Sahara desert is distributed into a pool of many glass houses. Three pools at least varying in depth are established in the glass house sealed up, and the last pool does it with convenience, making it for dry collection of the seawater residual substances. The seawater introduced into the glass house gradually concentrates by being heated with the sun, but the steam which was released is condensed dew in a glass wall surface cooled through the night and changed to liquid during phase transition, thus the liquid obtained is gathered in the big underground pool. The big difference of the temperature between day and night in the desert should bring switch to fresh water of the seawater by extremely high efficiency.

Thus, a large quantity of seawater which continues being drawn from the sea may supply further fresh water quantity while laying residual substances of enormous quantity in the central part of the Sahara desert. Vast energy is necessary to scoop enough seawater through the pipes and send it without depending on fossil fuels, but all of them should be able to provide by returning to electricity from sunlight through many sun panels which are installed in a very large desert.

As for the state of the community there, the artificial oasis city where a rule is accordingly will be born in each place in the desert corresponding on fresh water productivity capacity from the beginning. Even if the facilities for a meeting and education are necessary there, the facility is unnecessary. Rather I imagine when we will invite dignity of the dead person and time when we cannot but think about how we should face each other again in the far-off future. Because of my personal prediction, it is necessity that the times to be troubled with energy to burn up the human body causing death will occur before long. At that time, it questions whether humans may choose a burial. I cannot enlarge the graveyard where we stop cutting down the forest, but I will cry each time. Judging from the above-mentioned analogy, only a way to have a descendant make use of one answer for biomass, and that is to return it naturally.

At first, we must definitely grasp a fresh water supplying capacity in each fresh water production site, and further estimate appropriate population capacity accordingly if humans will be able to live culturally. With that in mind, we should draw up the citizen-based town plan that can use some fresh water openings for drinking and washing as the core in the neighborhood in a water supply site effectively. The people have to farm the area above, to a certain level widely with presence to give a self-sufficient rate of food. Of course, the ward for the orchard and the forest for the people must be located. What we forget locates a residual substance disposal factory mainly on NaCl. It is a matter of course, we must assign the amusement places such as an administration place, a school, a store, a hospital, welfare for infants and old people, a residential area, a drainage, and a park.

The left system which is indispensable if such an artificial oasis city is really driven will be the maintenance of a power supply system. The thing that is important is not only to accumulate the electricity produced depending on the sun and/or the wind, but also to store as H_2 after electrolysis of H_2O for security of the energy. To that end, it will be means to cover the roof of all buildings and houses with sunlight generating panels and to use feces as a biomass.

In addition, high durable materials are demanded in all structures. However, the organic raw materials are destined to dry up and are a concern in the future that they will become too expensive to be adopted. For example, the surface painting with an organic agent is very effective to extend the service life of concrete construction, and then its use will be limited to special cases.

By the way, it has been said that a cause of the dilapidation in the Middle East and nearby eastern areas has the abandonment of the farmland by salt breeze damage. If so, on making a garden city which assumes an artificial oasis in the

Sahara Desert area, we will have to push forward measures to reduce damage by increasing salinity due to irrigation. To our regret, however, I did not find preventive measures against damage from salt breeze that I can appreciate. Three kinds of choices for preventing the damage by increasing salinity are thought about preventing a rise of salts. (a) The final coating and painting of the concrete surface for the time being. (b) While is cheap and oil products are available, the spread of organic membrane such as polyethylene on the layer under the ground. (c) Adopting both law of (a) and (b) for the time being, which we unify it after appearance of a more effective on salinity. However, for fruit trees we cannot but adopt (b), because the root system of woody plants grows in breaking through the frame of (b). In this case, 2 kinds of the plumbing systems are necessary to send air to the root system of the tree as well as sending fresh water including nutrients.

The crop plants cultivated as staple food in the area spreading out in the subtropical zone from the tropical zone are corn, rice and yams. However, these are the results that they had adapted to a photoperiodic condition, but the requirement of these plants for the degree of water supply is greatly different, for example, corn plants are relatively resistant to drying, and rice plants are extremely vulnerable to drying. Because the artificial oasis city will be able to make every environment concerned about fresh water with which this oasis city will cultivate all kinds of plants which have adapted to the region in latitude zone.

However, a big unsolved problem is still left. A part of

humans evolved from an anthropoid ape in North Africa, moved to the east before long, and arrived at the area of the Tigris-Euphrates River and let Mesopotamia Civilization under existence of fresh water and sunlight on this comfortable fertile earth, in which fresh water had been shared by irrigation for farming. The local latitude of this area was located in the subtropical high-pressure zone, being about the same with the Sahara, and resultantly the quantity of rainfall through a year was less than that of evaporation, thus causing desiccation. However, the irrigation causes a drop in food production and brings the end in the civilization before long, because it was accompanied by salt breeze damage. This fact suggests that the artificial oasis city in the central part of the Sahara Desert with irrigation agriculture does not last long because of the salt breeze damage.

In other words, it will be a suggestion that we should choose the Sahara which does not include salt in the underground stratum if we make an oasis city. A trial here is not talk by the extension of Mr. Nozawa's dream. By the critical situation that the earth may face in the future, it is the construction of demonstration city which is applicable to the ground of which the world is drying. For example, the introduction of this trial to the Oceania continent may be able to contribute to a reincarnation of the new life in the Australian Continent. Besides, I would be happy if it becomes an opportunity for new city planning with halo-tolerance in the times when humans will not be able to use the organic matter products cheaply before long. Besides, my dream establishes the oasis village in each place radially from there, and will begin to walk it, making a satellite-shaped society of a future design with a green expanse.

Finally, I will describe the whole system which I thought of, a system supplying seawater to the inland, and let's finish this item. The difficult problem in which is common to many desert areas is that there are altitude differences between coastal zones and many desert areas. Therefore, in order to supply the seawater taken in the coastal zone into the higher inland areas, some kind of pressurization system is necessary other than the laying of a pipeline made with stainless steel. In any case at first, the seawater must be taken using a suction-type pump from seas and stored in the primary reservoirs to remove the mixed sands, the soil particles and the pollutants mixed in it only for a while. But at this time, the numbers of the primary reserve pool for the seawater must be increased largely to eliminate by precipitating especially the soil and/or sand particles. In this order, the reserve pool which shows high transparency, the top liquid layer of the seawater is moved to the final water tank through a pipe with a filter layer to remove floating planktons lastly and finally sent to the pipe made with stainless steel for water supply through some type of water pump.

The realization of Mr. Nozawa's dream in a very large desert area becomes possible only when the enormous quantity of seawater is sent to the inland. To that end, the stainless steel pipe with a big diameter must be buried from the coastal place to the inland. Since the vast desert often opens in the area of high altitude generally, the supply of seawater from the coastal place to the plateau which is high in several hundred meters cannot be solved only by setting one pump of pressurization type in the coastal place. In order to develop the desert area on higher grounds, therefore, the progressive setting of many pressurization type pumps must be performed between the coastal place and the desired desert to conduct the seawater which is cleaned and stored at the shore. The setting place of each pump would grow in an oasis town because fresh water is usable.

By the way, the supply of seawater on the development of wasteland or desert where it is located near the shore will be enabled without being concerned with the underground laying of the aqueduct, because its diameter is small. Seawater supply to the top of the hills may be possible with the use of suction-type pump. But, also the small pump of a pressurization type should be adopted for application to every case. Here, the formation of the oasis village in the coast neighborhood could be settled with a cheap introduction system of the seawater, in which the thin pipelines are located in the air like an electric wire and a pressurization type pump is set in many tower buildings at each node linking them. The seawater pushed up on the tower will stream down the next tower with a natural pitch difference.

The tower for seawater supply must be constructed as a building having at least 3 floors. Most of the first floor is used as the pool in which alien substances and contaminants within the seawater are again precipitated and eliminated. The second floor is used for the setting of 2 type pumps. One pump is for the tank on the 2nd floor from the 1st floor and the second pump to the 3rd floor pool from the 2nd floor. On the second floor, a tank is also set in which small soil particles are again eliminated after precipitation. The 3rd floor becomes the roof, on which 2 tanks are set. They are connected with pipes to each other at the lower part and the bottom of one tank which is bound with the stainless pipe for seawater supply. However, the setting height of the 3rd floor is decided by regulating distance between the 2nd and 3rd floors conveniently for natural flow of seawater to the next joint tray.

In addition, the operation of each connection tower is unnecessary for the placement of the staff except the starting point and the last spot, because photovoltaic panels and batteries are located in the neighborhood except both ends. Moreover, the size of the connection pool and the thickness of the pipe may be decided according to the quantity of demand by fresh water in the oasis city. And the reason for an adopted tower method by laying of the pipeline, there was the security of a long pipeline and convenience for check repair.

Above all saying that human beings lived in the desert, the important drinking water was secured. Next, food is necessary. In the deserts located in the lower latitude areas, however, any plants may not be cultivated. We had to cultivate it not to cause damage from salt breeze even if a little water blurred it while controlling a rise of the groundwater by a capillary phenomenon. The method to hold damage from the salt breeze in check in the future when the raw materials of the organic matter products dry up is limited. Still there seems to be no method to coat the ground with concrete at depth of approximately 1 m below ground on the local entire surface, planning it as the cultivated area if it is cheap and keeps damage from salt breeze.

The concrete itself also has life and never goes like a natural rock in modern chemistry. What desert sand we poured over after having coated the concrete entire surface using a surface painting agent such as the long-lived polyethylene looked good, if it was cheap and usable as an organic matter in the future. However, even if the brush coating by hand does not mind, the use of silicate type surface reagent is valuable to coat the entire surface of very large concrete, which exists in large quantities in the face of the earth. If the artificial oasis town planning which I gave here deserves consideration, rather the concrete entire surface of the allotment place for the very large cultivation should be painted by hand with a silica salt agent.

The fact means that the large size heavy agricultural machines are not usable in this new oasis town in which I couldn't but coat a basement with a thin concrete for prevention of damage from salt breeze in this place. Therefore, if I surround even the side of the farm with concrete, the crops here will not change with Nozawa's origin in case of "a tomato tree". It is only a difference growing a tree whether it can grow in the soil or water. As mentioned above, his thought was that only Middle Eastern people could eat fresh vegetables in their daily lives, using a little fresh water well. In other words, it was the investigation of the cultivation method that we could give a maximum crop with just a quantity of minimum fresh water.

The important thing is to understand the principles that the growing power in the upper parts in the same plant reflects the vitality of the subterranean part which depends on O_2 respiration and CO_2 fixation occurring irrespectively to light. Therefore, it was only the cultivation method that not only, we dip the root system of plants into the nutrient solutions but also we continue to bubble it with O_2 and CO_2 during a cultivation period, thus the tomato grew into a tree (see photo: CO_2 is stored as malic acid after fixed by the root system of all kinds of plants, which is then converted to other organic substances).

Thus, the food production that is going to be carried out in this new oasis town does not depend on existing agriculture anymore. The food production that all processes are controlled based on information is industrial and, except a small portion, must bring tomorrow's humans to the main industry to partake in environmental regeneration and to supply biomass. To that end, at first humans must prepare a thin Stainless pipe of L-shape at both ends in large quantities, which innumerably has small holes. For my thought, those L-shaped pipes are spread over the concrete bottom by furrow width, then covered with rough sand on the top approximately 15 cm, and finally backfilled with the earth for farming. But, as for the size of the part of the L character, around 1.2m will be proper because the L-shaped part complies with both sidewalls and line up, and the part of the edge appears on the ground. The air including both O_2 and CO_2 is sent underground through this pipe under higher pressure as needed. Thus, we always maintain work of the CO_2 reduction and fixation in the root system of higher plants in all farmland by being able to keep O_2 level in the soil high. A compressor operates automatically if the O_2 level becomes lower than a level with O_2 at several points under the ground and causes it to send air.

It is identical about the water supply. It has many methods, but, depending on the preference of each crop, the fresh water or the water containing some nutrients and minerals besides salt are supplied automatically by a pipeline spreading from a water tank under the ground, it goes without saying. The moisture in many points of the farms are measured and applied to controlling water supply. Only temperature is not controlled during the day time and night time in the artificial oasis town, being different from the plant factory during the modern fashion. The situation in the soil of the liquid fertilizer including the nutrient salts is always transmitted as information and adds the deficit all at once, and can adjust automatically. Thus, even if the food production in this oasis town is done with corn, yams or cassava, it is not established agriculture. In other words, industry performed here can say that it is food production business, environmental control business, and biomass production business, not agriculture.

Regardless of the supply method of seawater, the realization of Mr. Nozawa's dream is accompanied by the concentration of seawater. Therefore, the inorganic salts included in the seawater begin to crystalize and then deposit as residual substances. Most components are salt with Cl^- or $SO_4^=$, but various important kinds of rare metal elements such as Li, etc. are included, too, which may give opportunities to promote the inorganic chemistry company in addition to agricultural production in the area. Even if the demand for rare metals increases, however, in that time the rise in their prices will not seem to increase competition. The seawater residuals should be returned to the sea after dry solidification without saving any of it, if the location requirements are not blessed. However, we must not forget that seawater residues are a gold mine. For example, Li is not only necessary for storing the sunlight as electricity. Moreover, if the generation by a nuclear reactor could succeed in the future, the demand for Li would be superior to gold, because Li could produce tritium playing a solar substitute.

By the way, local peace is indispensable to make use of Mr. Nozawa's dream. Fortunately, the news of an intense civil war is not conveyed in every North African country along the Mediterranean Sea coast. Such a peaceful society condition suggests a possibility to be able to make use of his dream for the development of the Sahara Desert. If the countries of North Africa would accept his dream, they could produce high employment and might raise the population capacity in themselves. As a result, it is thought that his dream could contribute to the relaxation of the social strain in the EU countries by blocking the immigrants there.

(B) Introduction of rice fields contributes to regeneration of Earth's environment

If people put an important point to regenerate the Sahara Desert, excessive energy is necessary for water conversion before limited time. As for this reason, the very large rice fields may evaporate enormous steam in the atmosphere as they establish a big lake in the central part of the Sahara. It may increase frequency to bring the unlikelihood of rain by discharging a large quantity of steam in the continent inside. Still, the maintenance of the oasis town for rice growing is hurried firstly. The main premise needs maintenance only in the desert. When the service pipe from an underground of fresh water service pool considers tolerant to the oxidation and to the chloride corrosion, even if it is expensive, a stainless steel iron pipe should be used.

Furthermore, there are some cultivars of Asian rice which had lost completely the photoperiodic responsibility for flowering, which shows that rice grains could be harvested 3 times through the year in the subtropical zone such as the Sahara. In addition, many kinds of the Asian rice cultivars are aquatic plants, being able to grow in the fresh water including nutrients. However, O_2 is necessary for the growth of rice plants. For getting a rich crop harvest, farmers change the water level of the rice fields and further couldn't but dry the soil. In this way, they kept a body of rice plants warm and gave O_2 to the root system, in order to secure healthy growth of the rice by dry ventilation of the soil.

At this point, the big concrete box with air delivery pipes for Dry field farming is absolutely useful, and the water level adjustment in the rice fields becomes unnecessary in this oasis town. However, the rice cultivation method in Japan accomplishes progress for automation depending on work force lacking rapidly. The rice seeds coated with $CaSO_4 \cdot 1/2H_2O$ including iron powder are directly sowed in a flooding rice field.

Nevertheless, they couldn't cultivate rice plants without raising and lowering the surface of the water level for a feeding damage check, by rapid drop of temperature in the night time, by rapid drop of O_2 contact throughout the day, and by feeding damage due to a storm. However, the management of warm water is unnecessary in this oasis located in the subtropical zone and isolated virtually by other ecosystems.

All the management of water, supplement of nutrients, and sowing of new Fe-coated seeds is automated, and a pole premature variety of rice is developed. Moreover, the temperature difference between day and night is very effective to accumulate the carbohydrates and the lower temperature in the night time that is inconvenient for a high temperature obstacle, which brings the production of high quality rice. Further, there is no uneasiness of various feeding damages in the isolated oasis town. In southernmost Okinawa in Japan, there is a kind to finish rice variety from seeding to harvest in approximately 4 months which is already adopted. If such a rice variety will be used, it implies that the rice cultivation in the rice fields may greatly raise the population capacity per unit area because of production 3 times per year, suggesting even with the increase in population capacity.

The farmer who was already tied up and worked in a season becomes extinct. In the new oasis town, and in the neighborhood, the youth who can shift and change full use of science information there, may operate an associated apparatus, and may engage in food production business through the year.

Anyway, it is the same thing that the area flooded mostly is assigned for rice fields through the year as a newly artificial lake is made in the central part of the Sahara Desert. The steam which is born from the artificial rice field is expected to receive severe sunlight and have some kind of influence on the local weather. In other words, it may be said that the forest creation is unnecessary in the Sahara Desert. In the new oasis town, the rice plants keep CO_2 alive as starch and emit steam in the atmosphere like the woody plants, which retain CO_2 as cellulose and lignin. Accordingly, the people work here in food production industry, but also engage in the new environment regenerating industry to be classified as climate recall into the field of future vision.

Naturally, life drainage and agricultural drainage including various wastes will be necessary also in this oasis town. In addition, it will be rarely blessed with the rain despite desert. It is natural that they make use of another route in the recycling society. The life drainage should collect into a tank fermenter, and the rainwater and the agricultural drainage in place should be utilized after collecting in the hollow nearby. The former should drain the residual substance liquid into the rainwater pond after having made use in methane fermentation. The liquids collected in the pond can again make use for the production of fresh water.

By the way, the place of oasis in the Sahara Desert which I choose without any permission is located somewhere in the Algerian country. If so, it may be said that the seawater is drawn under the understanding of the Algerian government from the

Mediterranean Sea which is even slightly high in rational seawater temperatures. But what assumed an intake on the Atlantic side of Morocco daringly here is because I prayed this plan for global warming checks for an international action being targeted by sitting astride two countries.

(C) Meteorological improvement of Eurasia Continent holding earth's fate

In the end of December, 2014, some Japanese newspapers reported that the Chinese construction company which undertook construction of the great canal that bound the Pacific and the Atlantic planned by Nicaragua government had really begun to dig. Not only Panama Canal pushed forward repair of the widening, but also Arctic passage had already entered the year before last for a stage of practical use, about 30 years earlier than expected as a result of progress of global warming, guessed that a drop of the demand canceled the predicted Nicaraguan canal. When northeast Asian countries trade with EU, the economic route is through the North Pole Sea, although it is limited in a season. Then, the better route is through the Panama Canal, and the worst one is the Suez Canal.

The Chinese government understands the above merit and works on realization of the Arctic passage at the very beginning in cooperation with Russia. In fact, China even has already begun an examination service from the year before last. To see the result, many shipping company lines began to think about distribution through the Arctic passage. The problem is why the Chinese government understands that Nicaraguan canal but the Chinese company does not participate to its construction while financial support to Nicaraguan government is provided. Furthermore, the plan of the Nicaraguan canal makes use of Lago Nicaragua, which has an ecosystem peculiar to the subtropical freshwater lake. It is a matter of course, but the neighborhood protested movement against the canal construction which is generated to keep the valuable ecosystem. From the viewpoint that affected Biodiversity Convention, I cannot also agree to the canal construction that may disturb the valuable ecosystem of a subtropical freshwater lake. Nevertheless, the construction began. It seems that the Chinese government has some expectations.

Probably, as for the Han race becoming the leading role constituting China, it is thought that it originated by the group which remained being possessed by the existence of abundant fresh water around the Yellow River and the Yangtze River, thus both big rivers with chance on the occasion of ancient movement of a Mongolian spot and the bound humans. But a group with the primitive Han race avoids the Yellow river flooding in the lower basin before long, and begins to demand a home in the upper reach regions. As a result, some Han race will have a child of mixed racial origins between a west to a white of Central Asia, which the mixed bringing to hybrid vigor. Thus the Han group could get an advantageous breeding, and moreover, could live in a large area blessed with fresh water supply under the existence of two big rivers. In the times when the weather was mild and still wet, it is estimated that most of the Han race lived in the basin on the Yellow River. As for this estimation, it seems correct that the capitals of the ancient Chinese nations were located in the upper area of the Yellow River, such as HaoXI'an and Luoyang, as well as the remains mentioned previously.

In China, there are two big rivers of the northern Yellow and the southern Yangtze, where the latter spreads through the good mild climatic zone. Nevertheless, the civilization and the economy prospered mainly on the former. It is thought that one mere reason that established the nation's capital in the upper Yellow River will be to say that it was easy to trade with people of Central Asia for ensuring safety. The course that performed it was the Silk Road, and firstly the structure which supported the security was the Great Wall of China which I wrote. And deforestation for the construction was a motive of the desertification of China.

My understanding about such ancient times in China does not seem to be wrong from recent DNA analysis. While the people of the Han race are blessed with abundant fresh water of two big rivers of such ancient Yellow River and Yangtze River and are full of prosperity, they seemed to be gradually stained with Sino Centrism that may despite the neighborhood. But it is thought that there is some background to drive this feeling to real action, there seems to be a drop of the population capacity due to desertification flocking from the continent inside. China is a country having a very large area, but the area that green accounts at a color aerial photograph on Google is only about half in which mountainous areas are everywhere included and the cultivated area is not known as to how much land.

According to FAO 2003 report of the world population, the self- sufficient rate of food has begun to cut 100%, too, with a population of approximately 1,350 million of about 1/5 of the world population. In that point, the Chinese one-child policy is highly appreciated, but one of supremacy principle of the false Sino Centrism in the South China Sea is not acceptable. For example, the Chinese government has pulled a military doctrine for its own country without permission under the pretense of First island chain from right above the Japanese islands, the Philippines and Vietnam, and now is making the population island with the airport without permission from the Philippines and Vietnam in the South China Sea, preferably, I want China to show the posture as the large country so as to entrust the artificial island to international joint management.

When the Nicaraguan Canal will be completed, the global warming advances more, and the Arctic passage must become the regular line for shipping except the short term with the winter season. The shipping companies of Japan and Korea must use the North Pole route for EU, except the winter season during which they would use the Panama Canal. The shipping company of China will be the same, too.

Strangely, this canalization plan had been pushed forward between the Nicaraguan government and some Chinese companies in June, 2013, but the Chinese government had wrestled for the use of the North Pole route positively in those days and had succeeded to navigate a containership in the summer in that year. In the autumn of 2013, furthermore, the

Chinese government made a foundation plan of the AIIB and called participation in each country. Judging from this news in 2013, the Chinese government understood that a construction plan of the Nicaraguan Canal did not balance with investment enough and recommended that Chinese companies undertook the plan and would manage to roll up each country to support it financially. Although I have understood Chinese Sino Centrism, it is difficult to imagine that the Chinese government might not be really concerned with the plan that lacked in such rational possibly.

While the Chinese government ratified the Biodiversity Convention, I could not agree to the Nicaraguan canal plan, because it ignored the opposition movement of the local conservation of nature groups, and it competes with the expansion program of the existing Panama Canal. Furthermore, the Chinese government may involve other countries financing AIIB. In order to support the development of Asia, Asian Development Bank had already existed, and a fund was a story to finish it if the capital was increased for the new action that had been pointed out here. Fortunately, Japan did not meet this request with a wise decision.

I think that China should take action to get back to the environmental disruption that their ancestors have caused. Even if the help to the Nicaraguan canal will cause more destruction to the global environment, there is no function for the improvement of tomorrow's global environment. There will be the role that the Chinese government should check the desertification of its own country not tormenting tomorrow's Nicaraguan nation by ridiculous construction. Hopefully, China will support the regeneration of nature in Central Asia where ancient China had traded flourishingly.

The elements which human survivalists could not miss have been arranged with sunlight and fresh water, here. By the way, the damage that humans faced in global warming was now caused by the gross quantity of rain and snowfall being smaller than the gross quantity of the evaporation drying. The damage became the local desiccation, and attached a handle to the limited groundwater and, to the barren land, in sequence to the desert. According to the future prediction of IPPC (Climate Change 2013), the most serious area damaged on earth by global warming was located in the Eurasian Continent inland including the Chinese inland. It seems that the most serious ground is the environmental dilapidation in the Central Asia area which are symbolized by the extinction of the Aral Sea which was produced from failure of the irrigation agro-politics by post-war Russia. This Aral Sea was a salt lake 4th place in the world until 1960s, but now parted into 4 lakes, and the total area is reduced 1/5, and a part already became desert in which salt began to blow on the surface of the earth. Now the salt spouting place was unworthy to use, except spreading over solar cell panels.

From the end of the 20th century before a tragic situation advances to here, a way of the Aral Sea regeneration has begun to be considered in various ways led by Russia. However, because this effort was opposed by the interest of each country were opposed and couldn't but do with the present correspondence, the environmental disruption of the Aral Sea is reduced by drying before feeling like now attracted international attention with the picture which could not cry. It is thought that, after all, the reality that the west area influenced the ancient civilization of China has finished drying and it has some kind of relations to Chinese desiccation. If so, the Chinese government should firstly change all the labor and fund that China was going to pour into the Nicaraguan canal construction plan to the regeneration of the Aral Sea at the very beginning.

(D) Regeneration of the Aral Sea begins from the level up of the Caspian Sea

Unfortunately, there is a difficult altitude difference in the Aral Sea and in the neighboring Caspian Sea. The difference becomes 50-60m at attitude, but the Caspian Sea lowers approximately 24-28m above sea level adversely. Therefore, if someone makes a canal between the Caspian Sea and Black Sea and if it sends seawater into the Caspian Sea from the Black Sea in order to make the surface level of two seas surface to the same, resultantly the altitude difference between the Aral Sea and the Caspian Sea will be shortened to around 25-29 m. If such a position has a relation, at first those 3 seas must be connected in some way.

In this area, it is predicted that the driest damage weighs heavy on global warming geographically as described previously, but it is the lands where the destruction of the global environment has been pushed forward artificially, such as the deforestation in ancient China and as the hyper-irrigation in Soviet Union Era. Now, two major large countries are concerned with it, responsibility is in the situation that there is as a permanent member of the United Nations. Accordingly, both Russia and China should make a regeneration and remodeling plan for Central Asia, and at first have a duty to appeal to each related country for participation. As a concerned country, there are as follows: Kazakhstan, Uzbekistan, Kyrgyzstan, Turkmenistan, Iran, and Azerbaijan. With that in mind, these countries play a key role to get financial cooperation from the global community and, not AIIB, establish the Central Asia Regeneration, Development Bank and should plan financing.

My idea is to directly connect the Caspian Sea with the Black Sea by constructing a great inland water way that enables the cross-purpose navigation of the large ships. Now, the Caspian Sea and the Black Sea are coupled through Volga-Don Canal and the navigation of the ship is enabled by the opening of a gate of 13 in total with a pitch difference, but this system which ignores a pitch difference does not contribute to restore of the earth. It seems to be an unavoidable plan judging from a former engineering technology. It is to combine the Caspian Sea and Black Sea at the same level to be demanded for tomorrow. What I do not forget about Russians here is that the responsibility that accelerated desiccation of Central Asia may be caused by failure of the agro-politics in the former Soviet Union times.

So, the persons in charge must get the understanding of

people who must move out when the sea surface of the Caspian Sea increases 25-29 m. When the Caspian Sea is linked, the Black sea in the canal, the Caspian Sea area greatly spread through the west side from the north side of the Caspian Sea around about 35% greater, following probably eviction in some places. Moreover, all distances of the canal which link the Caspian Sea and the Black Sea passes in the same Russia and often passes an area of 0 m above sea level implies that the construction of this canal advances smoothly without a problem. Thus, the expanded Caspian Sea will add a role to raise steam content in the atmosphere in the continent inside.

Next, it is the regeneration of the Aral Sea with the original expanse and depth, but an altitude difference near 30 m is left between the Caspian Sea yet. In order to send the seawater of the Caspian Sea to the Aral Sea, it is thought that 2 or 3 big reservoirs which are established in any lot between two seas are necessary to get over more than the attitude difference near 30m. In each reservoir, two kinds of water pumps of a pressurization type and an absorbing type are installed pairwise and kept alive for the regeneration of the Aral Sea by the full operation of both. The reason to set a double pumping system in each place is because the water supplying time is limited to a daytime for depending on the photovoltaic power generation system. In addition, an outlet in each reservoir is established for discharge of soil and sand accumulated, and for drainage of torrential rainfall.

By the way, all of driving force of the system here depends on the photovoltaic power generation, but the accumulation of some electricity which is produced during the daytime is demanded to plan the electromotive averaging of the night and day. To that end, it is desirable to attach the above mentioned PSH generation facilities to each water reservoir as well as to the Aral Sea. Accordingly, the water level of the Aral Sea assumed in principle in the regeneration of the original sea area, but, as an operation share of PSH to introduce for the stabilization of the electricity circumstances in the area, the Aral Sea utilizes a 30 m high position as the adjustment water level at a higher water level.

(E) Contribution of the regenerated Aral Sea to the breakthrough of earth's crisis

The water conversion of the seawater adopts the technique which is the same as that is used in the Sahara Desert, but it is only difficult for this distinct to find strong sunlight quantity in temperature zone area through the year. On the occasion of the seawater to the fresh water conversion in this area, therefore, a work raising seawater temperature with a heater using accumulation of photovoltaic electricity or using electricity from PHS in the seawater pool in the nighttime is necessary. The large quantity of fresh water provided for the regeneration of the Aral Sea area merely loaded work to restraint of not only the past eco-systematic reconstruction but also the global warming with expectation. Therefore, the provided fresh water should be used for forest creation, from the stage with the prospect that the self-sufficient rate of food in the area can serve. However, the planting of woody trees is not possible in the place of damage from salt breeze. To that end, the salts attaching to those local soils and sands must be washed out using a large amount of fresh water.

The big fresh water demand in such a midcontinent is accompanied by the sedimentation of a large quantity of inorganic salts adversely. As a result, all kinds of inorganic salts contained in the seawater will accumulate around the Aral Sea necessarily. Total quantity of inorganic salts reaches 3.5% by the weight ratio really in seawater, so it should invite the company which assumes its raw materials around the Aral Sea. Residual salts of the seawater are mainly 77.9% sodium chloride, 9.6% magnesium chloride, 6.1% magnesium sulfate, 4.0% calcium sulfate, 2.1% potassium chloride, and include some rare metals. The situation that can make a definite promise of the power supply seems to be convenient to invite some inorganic chemistry industry using those salts as the raw materials.

According to IPCC (2007), the increase of forest area is necessary to increase the absorption source of CO_2 but it is reported there is not the land suitable for planting woody plants. The regeneration of the Aral Sea was not merely returned to the former condition and it was the plan that had an aim to establish the vast forest area in the Central Asia, which was a purpose to retain CO_2 which contained extra in the air into wood as biomass. If this forest appearing here is a complete artificial forest, we must do it with the biomass that all is easy to completely use for human beings. If the environmental conditions are good, the tree continues growing for several thousand years and can save CO_2 in materials. In addition, the tree covered the space to 20-30m above the ground with green unlike grassland and a shrub zone, and so we could take in CO_2 from 3-dimentional space of the surface of the earth, being able to reduce atmospheric CO_2 effectively. Since it is an artificial plantation, planting must be performed it under the plan to minimize a risk of outbreak of the large-scale forest fire and the spread of fire. Planted trees are a biomass, and if a forest fire will be caused, not only it is destroyed, but also it will exhaust CO_2 in the atmosphere at one sweep. To that end, we must make the forest which could be settled as far as there was the forest fire even if it was generated. The compartmentalization of the forest is made using the road with the water pipe, and a wide farming place or where grassland is inserted between the forest and the road. Under such a compartmentalization, the coexistence with forest management and agriculture or the stock raising industry can prevent spread of a forest fire. Humans acquire biomass by thinning, depending on time, and can be brought up in the full-scale vast forest area. The regeneration of the Aral Sea is for that purpose, the working of the artificial forest may increase opportunities of the rainfall and will take a role of mass fresh water supply.

The increase in atmospheric moisture content in Eurasia midcontinent due to the regeneration of the Aral Sea will be favorable for China. And, judging from the present conditions with a big income gap between the peoples in a coastal place and in an inland place, it will be able to welcome that the way to the opened for approaching relatively easily to the Caspian Sea from the inland of China. And, I have written that China should cancel an action for the construction of useless Nicaraguan Canal and should make an effort for the regeneration of the Aral Sea for the future of the earth and the tomorrow's profit of its own country. It is thought that there is one of the solutions of the local difference of the income in China for the development of the China inland area where it becomes a backyard.

By the way, I have focused in the social figure which we would meet after the drying up of the fossil fuels which I have described. In other words, this is how humans should make a soft-landing for that time. One thing depending on the fossil fuel the most is aviation now. Accordingly, whether or not H_2 gas is able to change the driving force of the airplane from the fossil fuels must be researched and developed further. But I do not think that the H_2 gas can really fly many airplanes as fuel, even if it is identified as a possibility. The raw material of the H_2 gas is water only, and it is not possible to use poor natural energy then to take out fuel for planes anymore. This possibility suggests that the day of the end times of aviation will come sometime soon. Global warming advances more at that time, and the Arctic Ocean may be it in the times to permit the navigation ships without freezing up through the year so it may disturb navigation in the winter season. Moreover, it may be the times when the Caspian Sea in the Eurasian Continent is linked to the open sea called the Atlantic.

According to IPCC 2013, on the other hand, the tragedy by the global warming at the end of the century is caused by a rise in sea level in the whole area on earth, but the desiccation will become particularly serious in the central part of Eurasia Continent, especially in the coastal areas from the Black Sea to the Aral Sea. The rise in sea level with global warming is caused by two elements; the first is a thermal expansion of marine itself and the second is an increase of the substantial water amount in the sea area caused by the melting of the eternal frozen ice. The quantity of sea water which is poured in the Aral Sea on the occasion of the regeneration seems to be understood as partial collection of the ice which melted. Therefore, it is thought to be fair as a strategy that the establishment of many fresh water retention facilities should be evaluated in Eastern Europe area where it may be hit by serious drying with global warming.

As mentioned above, in addition, if the times shift sometime later from the air plane toward shipping, it will not have to be a dream to establish the way for ships there. Fortunately, a network of the corridor of the water has been already made throughout Europe, and is. For example, even if we do not appear in the open sea from the Caspian Sea to the Black Sea, it is possible if we use Rhine-Marin-Danule Canal, White Sea-Baltic Sea Canal, and the Volga-Baltic Water Way. From the Black Sea, there are already two ways to the Baltic Sea from the Dnieper River in the Ukrainian; the first is through the Neman Canal in Lithuania, and the second is through the Bug River in Poland. The Volga River also goes north into Russia from the Caspian Sea and has a route toward the White Sea or a route toward the Baltic Sea through various kinds of canals.

10. Japan and China Images in the Future after the Drying up of Fossil Fuels

Before guessing the peaceful future of East Asia, at first, we must recognize the facts that two major economic powers exist in the area, which GDP calls China with second place and Japan third in the world. In addition, these two countries are guessed to be affected by different climate change in response to global warming, because China is a continent nation with a huge population, while Japan is a marine nation of the islands. In the Japanese Islands, besides, a steep alpine belt of 2,000m grade ranges from the south towards the north almost in a right angle for an axis of the earth. Therefore, the Japanese islands can catch not only the steam from the Pacific Ocean but also the steam from the Japan Sea, suggesting that Japanese would not meet eternally the serious water stress as far as the earth is a planet of water, even if they would increase a risk with a disaster by rainfall. In other words, human beings in the Japanese islands continue fortunately being endowed with the fresh water resources which are indispensable for human survival, although the area of the country is narrow and would be troubled by a volcano disaster and an earthquake disaster besides a flood disaster.

On the other hand, the People's Republic of China, a continent nation, is a large country with a population more than Japan 10 times, and moreover. It greatly spreads at long distances from east to west and from the subtropical zone to the temperate zone. Even in 2003 when global warming began, China boasted of a self-sufficient ratio of food that 100% were close to being like Asian country of India, though it is a continent nation (FAO. 2005). According to Epoch. Times Japan. (2013), however, the self-sufficient rate in China fell down to 90% afterwards for 10 years from 2003 and says that China must increase farming area of 200 million hectares to fill up 100%. Probably, this drop seems mainly due to an increase of Chinese population because approximately 6 million people increased even in 2014 (IMF). However, the story that this drop associates also with the expanse of useless land where approximately 16% of the existing farmland became unsuitable for farming by damage from salt breeze due to the underground-water dependent farming and by heavy metal pollution due to the progress of industrialization in China has a very serious problem. This fact suggests that the fresh water-supplying power in China is already coming to a limit.

Not only can humans miss out on fresh water in their lives, but also humanity is established with fresh water for agriculture and industry. According to WB (2001), all developed countries secured fresh water resources above a certain level, which is generally distributed 59% for industrial use, 11% for life water and the remaining 30% for agricultural water. In contrast, the fresh water in the developing countries is used differently. Most of fresh water 82% is used for agriculture, 10% for industry and only 8% for life. Probably, the communist party in China maintained dictatorship using its first condition by starving its people with limited fresh water resources. In fact, the annual precipitation per year in China is only 645mm which is less than half of 1,668mm in Japan (FAO, 2014). This is less rainfall than 715mm in USA where GDP is the number one in the world, which implies that China can never surpass the United States in GDP, particularly GDP per person.

In spite of it, the communist party political power succeeded to bring up China's GDP to second place in the world. This economic growth was a result that Chinese broke poor fresh water resources, in which the surface water was effectively used with the improvement of the irrigation canal net, the underground water was kept alive, the waste water was reused after purified, and sometimes the sea water was converted to the fresh water. However, the dependence on such fresh water has brought a negative result to be called serious environmental pollution in China, suggesting that the economic growth in China is coming to the limit. The reality that the fresh water quantity of natural resources which is indispensable to both the industrialization and life-improvement has already come to the limit implies that the economic growth in China cannot but slow down gradually and is coming to an end. Some people predict that China will achieve become the world's best GDP before long, but I think that it is only an impossible delusion unless they sacrifice the life of the Chinese people, because the quantity of evaporation in China of a continental nation will be superior to that of rainfall even if the global warming progresses further.

If so, what kind of action should Chinese themselves take for the existence of its own continent nation in a period before drying up fossil fuels from the earth? China of a huge continent nation is divided into the east and the west area by the mountains 2000m-3000m class from the Myanmar border through the west side of Beijing to the Siberia border, the line of which has been called as Heihe-Tengchong Line. The east side of this line catches the steam from the Pacific and is blessed with a rain and snowfall, where 94% ($1273X10^6$) of the total population live in spite of an area of only 36% (about 3470 km^2) of the country. Thus, the substantial population density in 2002 is about the same with Japan (about $338X10^4/km^2$) and China (about $365X10^4/km^2$), because the area of the east side is approximately 10 times larger. It may thus be said that the action demanded on China is to keep the very large drying area to occupy western 64% (about 6170 km^2) of the western country alive for the own future and for the future of the earth, since the population density of the east side is extremely high.

However, the western part of China which consists of the deserts and the barren lands at the higher altitudes becomes the plateau of the grassy plain where the trees are not fully grown because of a big temperature difference of the day and the night when the temperature shows a subzero occasionally even in summer. However, what we do not forget is a fact to say that the weather of such western area had been caused by the ignorant action of ancient people of the Han race in China. As described previously, the forest in the Chinese continent inland had been destroyed firstly on the construction of the Great Wall of China before A.D., and further collapsed by the Great Leap Forward of a stupid leader Mao in modern times. Not only the desiccation in the China inland became a hindrance to the economic growth of China itself, but also influenced on the weather of neighboring countries as sandstorm and/or dust storm. Since the weather of Japan and Korea had strongly been influenced by natural environments of China's inland, I remember that I agreed to the construction plan of the Three Gorges Dam in China which had been objected to destroy the neighborhood and river environments from the representative USA and a lot of others, at the times when I was concerned with the first Earth Summit. As for the reason why I was able to agree to it, a purpose of the construction of the dam tied running water of the Yangtze River by connecting with plural water lines to the Yellow River which often cuts off water supply, and supplemented the lack of fresh water in northern China. The construction of this dam let me expect that it could return the clear sky to the northern Japan from late fall to the early spring in the time of my boyhood. Now, two waterways link northern China and southern China, but any improvement with the global warming is not accepted in northern Japan either. Now, the sandstorm and the dust storm from China make a range of black ash on the outer layer of fresh snow.

The quantity of precipitation in the southern China area is supposed to increases more and more global warming as it advances, while the quantity of precipitation in the northern China area is not predicted to increase so much because the higher mountains in the Japanese islands will isolate from the steam of the Pacific, although it will somewhat increase for the gross amount through the year. In the future, therefore, the deficit of the fresh water in the northern China may be made up by the further increase of the fresh water supplying through the canals and/or pipes from the southern China, although there is no extra fresh water there either. The most important problem assigned to the Chinese people is how they will be able to regenerate the very large area of western 64% of Chinese continent for the future foreseen when desiccation advances with global warming more.

The future fate of China must be decided whether Chinese people can utilize very large about $700 X 10^3 km^2$ of the western part that had been ruined, the area of approximately 2 times of Japan. Now, China seems to submit its own future fate to marine advances. Not only the construction of Nicaraguan canal but also the construction of artificial islands within the exclusive economic zone of the Philippine only plays international dissonance, both not contributing anything to the future earth. China, a large country with a huge population takes historic responsibility for having exhausted a global environment by letting the past make half or more of the country drying. China should work on the regeneration of its own country in response to it. If the Chinese government changes the policy direction for the tomorrow's earth, neighboring country, Japan of the GDP third place should support it.

11. The Eternal Peace between Japan and China Related to the Fresh Water Supply

In 1990, the very beginning argument at the Earth Summit began with a prediction of the population capacity of the earth. It was, so to speak, a prediction of the amount of production in the future of cereals and potatoes which will become the staple food. In the estimation, the pickup of various kinds of factors that would affect it and an order charge account of the degree of those relations were necessary. The certain thing is that fossil fuels buried underground dried up sometime soon, and that most would be released in the atmosphere as CO_2 which absorbs the longer waves emitted from the sun. If it becomes so, anyone will think that the planet earth must be in a situation as in a greenhouse. In the greenhouse, the ice melts at the very beginning. Many members must have worried about even the possibility that the eternal ice on the earth including that on the Antarctic Continent finally melts, although the snowfall there still continues, and the seawater increases the volumes by making thermal expansion. So, on the estimation of population capacity of the future earth, how much the sea level rises and how much the land area shrinks will become extremely important.

Humans must work hard to avoid the predicted worst situation while we can use fossil fuels. The only method thought about is that some of the quantity of water eluting from the eternal ice of each place is to hold down an afflux into ocean by establishing a big artificial lake or sea. The appearance of a very large hydrosphere in the land may produce the new troposphere of the steam and may let a heat balance on the surface of the earth become dispersion, which can expect an effect to restrain the outbreak of the local abnormal weather. However, the setting of the larger artificial hydrosphere is not able to perform on the land having a high population density. A large amount of cost is necessary for the construction of the artificial hydrosphere, and so the setting must be well chosen for producing value in the neighborhood. I think the appropriate place would be in the western area of the Chinese Continent which had made the forest in the previous century. In other words, the big business which establishes some large hydrosphere in the highlands of an altitude of about 500m is the earth renovation in the Eurasia midcontinent, which does not seem to be possible only by the effort of China.

It is just the time when East Asia can build eternal peace, if Japan of the same cultural sphere sharing a Chinese character called a "kanji" holds out a hand to China at the very beginning. In that point, it is the same as the revival plan of former Aral Sea, which was planned other than a purpose to gain rainfall by restoring the steam troposphere in merely Central Asia to prevent even a little rise in sea level. The bright future does not open out in the People's Republic of China, even if it plans a marine advance at underground resources while troubling the other countries. Maybe, the thing which they will lose is much bigger than getting a temporary right by power, resultantly

China losing international dignity of the Han race that is the main Chinese race. Only when China can make the existence that they could utilize the earth on its own, can they be able to boast to the world.

The Chinese government did not notice that the supplying ability of fresh water resources is an important limiting factor by which the economic and social growth is controlled. Furthermore, the Chinese government seems to think less that the phenomena on the atmosphere of the PM.2.5 pollution and on the environmental pollution of hydrosphere takes place indirectly due to the poorness of fresh water resources altogether. When it looks at the news pictures delivered by the China, it is certain that the economic growth in China is completely prevented by the lack of fresh water resources now and is slowed down. Even if the Chinese government plays for time to obtain a large quantity of fossil fuels from the foreign territory, since they are limited, the day when it cannot but complain will come sometime soon. It is not ocean advance bringing strain in the neighboring countries that would like to ask the Chinese government by all means, and it is to declare a suit that China want to start the reconstruction of the dilapidation in the Chinese continent which the ancestor and the senior of oneself Chinese had done for the global community.

From here, I perform concrete suggestions so that the Chinese government can have possibilities to share a critical mind with me sometimes soon, or China may be interested in this article to some extent. Humans cannot walk on the earth in the future unless Chinese people show posture to navigate as the friend that a Chinese friend occupying 1/5 of the whole planet earth. If China leads it and embarks on the hydrosphere construction of the inland to protect many shorelines and harbors of all parts of the world, then this will follow to China giving an opportunity to act in the large country with other deserts. Maybe, the trial of China may be with a precedent laying the construction plan of the artificial lake to prevent a rise in sea level in the Australian and North American continents.

Fortunately, not only professionals of the water way construction is going to build the Nicaraguan canal but also enormous work forces engaged in the paths of rapid transit railway construction which will be finished and is predicted to be redundant soon in the China. The country reconstruction business in China will make use of their wisdom and experience, and it should be able to become a national business to produce much employment. There is no other country that provides such favorable conditions in any place other than China. Therefore, the Chinese should start reconstruction in western China for oneself seriously. The United Nations wants to lead it and support it and support it for the future earth like the Three Gorges Dam construction case. Neighboring country Japan should meet to a request with China, too, and do not forget about the Japanese existence in the Chinese government either.

Then, I want to introduce my own plan here. At first, in the places that are near to the fountainhead of two big Chinese rivers of Yangtze and Yellow which become together from a large number of branches, Chinese people pick by two places in each river, 4 places in total. On the occasion of choice, desirable place would be suitable for storing water as much as possible. Furthermore, it is also desirable that altitude is higher and ridgeline is near if possible, since it is easy to discharge the stored rainwater into the other side of the mountain range. If the places are decided, 4 reservoirs in total are built in each place. As in the case of regeneration of the Aral Sea mentioned previously, the rainwater is lifted up progressively to the reservoir set at the height of the ridgeline. However, when a shortage of water state happens in the lower basin of two big rivers, the principle that the stored rainwater goes back in the main stream is applied, and as a result, the water supply to the west side is stopped. The rainwater which was lifted to the reservoir established in the ridge and was stored there will then flow down to the destination in the western barren land located to approximately 500m above sea level through a water pipe line.

Based on the plan, a high difference approximately of an average about 1000m occurs born between the reservoir on the mountains and the hill area of Chinese interior. This great altitude difference will be created in 4 places, and be able to keep them alive in hydraulic power generation as energy of the position. For this, some reservoirs at a certain distance are established between the reservoirs and the destinations in order to set a hydroelectric power station. The significance that the barren land of the drying is moistened by rainwater is extremely big, while 4 places of great altitude differences are converted into electricity small as for the turbine of each hydroelectric power station being small. In addition, the water supply to the destination should be using a water supply pipeline as a general rule and should not adopt the water way method even if it is level ground, because salts in the soil dissolve into rainwater and the rain water will be partially lost by evaporation and by soaking into the soil. There is no meaning if it is not fresh water which is kept for drink use with the destination of the rainwater.

By the way, I have heard from the peoples who live as the neighbor of Beijing, a capital of China, are already troubled for water stress. On the other hand, the times began when Japan was already troubled by heavy snowfall in the winter season in the side of the Japan Sea. The villages, towns and cities were attacked by the heavy snowfall, paid money and labor in large quantities, and couldn't but struggle for a dumping place of the fresh snow. If there was not suitable vacant land that was suitable for snowfall disposal, each administration let groundwater gush out of the central part of the road for dissolving the snow and then drop it to the drainage of the road, or must carry it by car to the river. The heavy snow fall is natural disaster in Japan. Already in 2000, I have written the thought that the peace of the tomorrow's East Asia must be established by making the mutual two-way streets on a Japanese magazine (Seiron, 2000, No.4, 232-242), in which Japan lacked in natural energy resources and troubled for heavy snowfall is bound with China troubled with serious water stress. Because I was brought up with poor wartime supplies, I came to have the feelings that it was a waste to throw away into the sea when the fresh snow which had brought a natural disaster. On the other hand, I was thinking and I was sorry for the Chinese people who were in the struggle with water shortage that the suspension of the Yellow River water supply has already become common.

If China calls Japan for help in the future, at that time, the idea is the same way as that of Turkey, in which a ship pulls a big rubber bag with fresh water and sends it to North Africa. However, I did not think that Japan could relax their water stress at such a level to meet the huge population of China. As I could guess, the country which would suffer the most from thirst on the occasion of global warming would be China. It was thought at that time when Japan couldn't but wrestle with another mean neighbor that even a part of the rain snowfall which was drifted to the steep sea in the Japanese Islands constantly could come to supply China constantly. Of course, there is no help for it, if China says they do not need the cooperation of Japan because the Chinese government works on water conversion of the seawater. Depending on a method of the water conversion, however, I do not think that all demand of fresh water in the northern area of China can be met by only seawater conversion, because of the vast costs for energy consumption, for the update of the materials, and for the residual settlement after the conversion. If a superior leader appears in China, I think that he must get the touch from existence of fresh water discharged from rain snowfall in a neighboring country.

When fossil fuels dry up, I can imagine that the fresh water demand in the north Chinese district of the dense population area will fall into an extremely serious situation. The water stress in those area hold enormous population. The problem will not be solved only by any work of only its own country, when I think about the burden of the fresh water supply to the west side of the Chinese continent, the lack of beginning groundwater, a limit of the fresh water supply from the southern area in China, and a higher cost of the seawater conversion. And so, a leader appears in China who is going to open up tomorrow newly without being seized with the past history, and the times when he demands help from Japan throwing away a rain and snowfall in the sea may come. At that time, Japan may think about a supply system of the fresh water concretely if they possess it. This is because there is risk in Japan that the system in response to the request contributes to the cancellation of the domestic water shortage in Japan. Excepting a few microbes, all creatures can live during some period if blessed with only fresh water. The fresh water is so indispensable for the working of life and the homework that the leader should achieve is liberation from the water stress. I think that the economic development of China will not continue for a long time, because of the limit under the present water shortage. The maximum neck on maintaining the government by the communist party alliance autocracy is that the fresh water supply capacity is too small in comparison with population. Maybe, the present Chinese government cannot take a good balance among use of the fresh water for civic lives, for agriculture use of fresh water for civic lives, for agriculture use, and for industrial use, since the lack of the absolutely needed quantity of fresh water has already begun. There is no way to serve the demand balanced for fresh water supply proportional to population except that it increases the capacity of its supply.

When realizing it, the Chinese government must choose a commensal way with Japan which is an aquatic resource country.

It is certain that the times of the fossil fuels are over by all means sometimes soon. The forthcoming energy at the movement state is either H_2 gas or electricity which is saved in a battery, but H_2 gas can be made if there is electricity. In any case, it may be said that Japan is the greatest nation of rich natural resources if the raw material of H_2 gas exists in fresh water approximately unlimitedly. However, even if Japan is blessed with abundant fresh water resources before, it has been worried by a useless treasure without being endowed in electricity derived from a natural energy necessary to electrolyze it.

In fact, Japan is a narrow no resource-rich nation, but fortunately it is an island and resultantly the territorial water areas are very large, which seems to have high potential if I assume it is a place receiving natural energy from the sun. However, it was natural that I could not make use of the whole territorial waters in photovoltaic power generation and wind-generated electricity, if the territorial waters were places of the life of fishermen and they were some leading figures with a self-sufficient rate of food. Besides I was not able to estimate the change of the solar energy capture rate by the rise in seawater temperature with the global warming that a steam pressure could come on the sea and could hold it.

Under such judgment, I was in a state of mutual reciprocity with Japan and China in the future and I prayed for eternal peace of East Asia and wrote an article in the previous Japanese journal. As for the contents, Japan sends fresh water through several pipelines to China, and on the contrary, Japan receives the electricity derived from the light of the sun in the inland of China. Japan had already made use of an undersea tunnel in the country. The Japanese society had already kept the undersea tunnels alive, and it did not think much about the difficulty to bind together by laying some pipelines on the bottom of the East China Sea because it was not very deep. When it would be completed, I thought that a relation of the give & take between Japan and China could be established for the eternal peace in East Asia, in which Japan supplied a large quantity of fresh water to China through the pipelines, and Japan received the large quantity of electricity from China through the cables in return. This relationship is the existence that they should maintain for Asia so that a new natural energy-dependent generation system will be developed.

The result of the research and development of the big company in Japan becomes recent news. Because that company succeeded in a trial to transmit the electricity generated on the sea to the land by a microwave, the practical use of the space-photovoltaic already disappeared in the dream. In a world of advanced technology, that company suggests that one step got closer to the development of the system which can transmit electricity to the ground point that it aimed at by wireless from space. In Japan, the light from the sun is always converted into electricity without the distinction of the night and day, and then the electricity will be transmitted to the ground by radio soon. It suggests the arrival of the times when the condenser is not necessarily indispensable for the stable supply of electric power, and further suggests that the submarine cable may be unnecessary for the transmission from China to Japan in the future of the fossil fuels drying up. Therefore, it will get off with only a water line system for China from Japan. The need of the laying necessary several pipelines for fresh water supply from Japan to China do not change. Rather, further research and development about the materials of the water pipes for laying may be necessary, because the rise in seawater temperature due to global warming would quicken the deterioration speed of the pipes.

As the People's Republic of China is the very large nation which ranges from the subtropical zone to the subarctic zone, I can suppose the country structure has some kind of influences on the future global environment. However, the Chinese government moves for the establishment of AIIB for securing rights and interests rather than the making of its own country. Looking from the side, at a time when the Chinese must make countries for tomorrow, I feel that they are swung around in Sino Centrism and I am ashamed. I cannot understand why the Chinese government brings the headquarter of AIIB into the suburbs of Beijing having been troubled with serious water shortage and why it makes light of intensification of expected water shortage. In my opinion, if there is a little consciousness as a member of humans, China which has 1,300 million or more and has very large land of drying occurred because of the failure of their seniors, should carry out responsibility to regenerate its own country. They could not understand the importance of the forest in the working of nature, and resultantly the Chinese inland had become a plateau of dry dilapidation. The Chinese government should recognize it just to return the very large ruined plateau back to green land, when China would contribute to saving a crisis of the earth due to a rise in sea level by global warming. Certainly, the action of Chinese to return fresh water to the inland must lower the risk due to rising sea level, because the area of Chinese inland is very large. Therefore, what the Chinese government should now make is Chinese Infrastructure Investment Bank and not AIIB. If the Chinese government demands a bailout for own regeneration, Japans should show the posture from the very beginning. By supplying the fresh water in large quantities, Japan can cooperate with the regeneration of China to account for a big area at the surface of the earth.

Some small marine nations are on the verge of the submergence with global warming. Moreover, this crisis is not a thing of the degree that the residents of the islands can evade by the construction of seawall to prevent the submergence from high tide or by the construction of a new artificial island. The only evasion means is to reduce quantity of water to begin to melt from eternal ice, and to flow into the ocean. That action can act without pouring a rain snowfall into the sea by keeping it alive in the land. If China understands cooperation with an equal action for a human community, the Japanese will save river water with pleasure and will work out a concrete plan to supply China with water. And probably, it is an opportunity that will contribute to the laying of the main water pipeline witch can be equivalent to a local drought disaster in Japan and the construction of the new net of the ramification from it. By that time, even if rainwater falls on roads or on roofs, it is merely assorted with life waste water and is incorporated in a net of the water and may be going to cross the sea to the China.

Preferably, China wants Japan to achieve a voice for the support of the fresh water supply for the making of country which will fit each other in the times after the fossil fuels drying up. If fossil fuels seem to dry up soon, probably China could not but call for help in Japan having abundant fresh water resources. However, then it is too late. While the prices of heavy oil and natural gas are low, the Chinese government must do decision for regeneration of the own inland, and if China want to avoid having a dispute between a neighboring country, they should act in order to demand the supply of fresh water for regenerating their own country from Japan as soon as possible. In my prediction, it is thought that oneself Japanese will enter to plan the partition method of fresh water resources within the country for the next development of Japan at that point for meteorological localization with global warming. Maybe, Japan at that time will begin to think about how people can concretely utilize a domestic rain snowfall. Before that time, if Chinese do not demand the support by fresh water supply, their national dream to be able to boast of in per person GDP will be cut off because of the absolute lack of fresh water resources, even if they struggle for it. By my estimate, the abnormal weather by global warming causes a local drought even within the marine country with higher mountains such as Japan. If abnormal weather will become common, I think that the Japanese government could not but plan about the construction of a fresh water supplying system for the further development in Japan. Before Japanese government will begin drafting a plan, the Chinese government wants Japan to call for fresh water support if it is possible. The Japanese would be happy if a main pipeline of the fresh water supply runs from Hokkaido to Okinawa and the network consisting of branch lines are completed in Japan by making use of a support promise of fresh water supply for China as a chance, like an existing railroad line net, an high way road net and an electricity distribution net.

12. Remarks in Front to COP 21 Beginning Soon

The UN has come as the International Day of Forests in the Vernal Equinox Day of March 21. As for the UN having but arranged such a memorial day, there would be a wish that it was warned to rapid decrease of the forest cover which went ahead depending on a rapid increase in the world population after the Industrial Revolution. The biggest reason that the UN couldn't but do so was because the forest was a leading figure of the fresh water supply for survival of humans.

As before, a fresh water supply power felt that it was with a maximum element for human survival and the importance did not change until the sun burnt out sometime. If humans could do it before the drying up of the fuel fossils, it was a wish that the human could erase a yellowish brown color from a picture of the planet earth which had been sent from satellite. In other words, an effort that humans played, barren land of the drying in the forest not only increased the fresh water resources but also decreased the atmospheric CO_2 partially. Of course, the quantity of CO_2 that the forest can save as biomass will be limited to a small portion of exhausted CO_2 by fossil fuels buried in the earth. Therefore, humans must harry the development study of the other CO_2 storage method to control the CO_2 concentration in a level in which CO_2 does not disturb the oxygen breathing in all creatures and it does not cause serious global warming. The study of the various kinds of CO_2 retention method has been pushed forward, but it is desirable to just purify it without dissociating CO_2, and to reduce the volume now. In that department, it will be desirable that the CO_2 liquefies in the bottom of the deep sea after having been purified, cooled and pressurized. As one way for saving the biomass, it will be possible that a small portion of the CO_2 accumulates as polymer graphite only for C element, which is useful as a building material.

However, global warming has given me a true experience to me by the change of a lot of imminent many natural phenomena, but the most serious situation has exposed the small island of some to a crisis of the submergence by natural disaster to be caused by a rise in the sea level. It is concerned with the factors that not only the fusion of eternal ice, the marine thermal expansion, the acidification of seawater, but also a future sedimentation of liquefaction CO_2 in the sea bottom. Not only that, the frequency of very large tropical cyclones in the category 5 has increased and caused many casualties, such as Hurricane Katrina 2005, Hurricane Sandy 2012, Typhoon Haiyan 2013, and Cyclone Botswana 2015. These facts suggest that action assigned to humans to wrestle at the head should be connected directly with even delaying a little in sea level rise.

In a sense, the act may be the divine message that human cooperates as an earthling while the fossil fuels are usable, and starts remodeling of the earth from the place where it can carry on in order. For it, vast governmental spending and work force are not necessary, and an international collaboration is indispensable. At present, it will be limited to the United States and Australia, two countries that the nation meets those conditions. Each country holds very large drying barren land in the alone in the country and can assign it for the construction of the fresh water reservoirs by own decision without coordinating the interest with the neighboring country in the continent inside. They can decide what kind of role they found, a lake to be expected in which area for the intention of the nation. Moreover, if it is the bailout that met national interest, the acceptance will be possible. The area of the drying in the North American Continent opens in Mexico, but the United States is the world's biggest GDP large country and it should earn time by preceding and moving it to practice.

All the areas of other barren land sit between many countries, and the UN cannot but take the role of coordinator there. Even in the barren land of the relatively small drying, there may be interest conflicts among many countries. In the Namibia Desert spreading through the southern part of the Africa Continent, the climate of South Africa, Namibia, Botswana and Angola are all involved with each other. Like the Namibia Desert which spreads out in the Atlantic west side, Peru, Chile, Bolivia and Argentina are involved with the west of the Pacific.

However, I think there is the area deciding tomorrow of the planet earth in the barren land of very large drying barren land to open North Africa along the Tropic of Cancer from the country inside in the eastern part of the Eurasian Continent. Therefore, the fate of the earth seems to be due to how the people who assume the area a home can cooperate with each other for earth-water preservation. With that in mind, the people who assume the area as home mobilize every possible energy to be opposed to the power of Horse Latitude developing force with global warming, and have no choice, but to bring in water of quantity superior to evaporating aquatic resources in the area. I worried about it, and I already sowed three ways on how to work on this article. The first is the indirect fresh water supply through the regeneration of the Aral Sea, the second is the water conversion of the seawater in the Sahara Desert, and the third is the fresh water supply by the water pipeline laying.

In the huge drying zone of North Africa from Central Asia, many races form each nation with their own different faith. Moreover, the most local people in this area are Muslims, and they send daily life mainly on the faith unrelated to science. In such a society where so-called Islam regulates everyday life, even the cooperation request to build the freshwater lake in the inland to prevent a rise in sea level will finish being born. Accordingly, the area that the UN first leads and calls for cooperation in the global community is the central part of the Eurasian Continent, particularly less than 0 m above sea level. But, I am extremely sorry that humans must wait for the conscious change of people in the Islam society having the potential that it is the most extensive, and can hold fresh water for preventing the rise in sea level in the land. But, the global community is not merely waiting a day to stand up because the Islam society itself breaks a crisis of the earth, and it should continue appealing for cooperation.

Therefore, the area where the UN can act on the interest adjustment in associated multilateral talks will be limited to

Central Asia for the time being. It may be in particular the help for human that the very large land which is almost near sterility of 0m above sea level exists in the area. In addition, if the day comes and it is able to link the area to the open sea adversely in a canal, it is thought that a thing to get is much bigger than losing something. If it asks why, the times of the plane are over sometimes soon, and this is because it cannot but recur in the times of the ship. But, the two countries of Turkmenistan and Uzbekistan may be going to lose the most of the country if it becomes so. Both countries may invite the times when they should work on a country structure to make use of the state of a new hydrosphere for new age. There should be the route leading to the open sea without keeping work as reconstruction of merely the Aral Sea. Great public works are necessary for it, but it should link the Black Sea, the Caspian Sea and the Aral Sea at the same sea surface. In modern engineering technology, it may link the Aral Sea at the same sea level through the Caspian Sea from the Black Sea not as a canal of Lock type. Fortunately, the company which completed the Bosporus strait channel tunnel was a Japanese company, which reports that the ebb and flow speed of this strait is relatively small. Even this large-scale construction will carry a large quantity of seawater in a continent, and it seems to contribute to the check of the rise in sea level due to global warming, in addition to the vision that a medium-sized ship will be able to navigate from the Aral Sea to the open Aegean Sea. The expanse of the sea area in Eurasia Continent inside plays a role to prevent a rise in sea level, and it is thought that the expanded hydrosphere lays a change for an area climate by giving birth to the troposphere of new steam responding to global warming. Of course, even if the local people do not chase a dream and merely carry seawater to each place with water pipelines from the regenerated Aral Sea and wrestle for the seawater conversion, they may spend wealthily just to have formed a new industry base, but they must always prepare for disposal of salts there as described previously. In any case, the introduction and expansion of a hydrosphere in Central Asia is a mission for humans who cannot but think with the problem that the UN plays a key role and should act for the practice.

In addition, the western Indian Rajasthan state is located under the Horse latitude zone and it is also with barren land of the drying. Accordingly, also the construction of the artificial lake in this area may work for the check of the rise in sea surface and may contribute more to the economy in India. However, there is no valuable area superior to the very large land of the drying of the whole Islam zone spreading from Pakistan through Near East to Morocco. For the survival of humans in future, it is a desirable story that the very large drying area such as the Sahara Desert where the Muslim lives would become useable by the method described previously in 10 (A) item. However, the liberation of Muslims from a strong doctrine of the Islam may be necessary to make use of this area for storing fresh water.

The bilateral relations that were seen in the North American Continent are seen between China and Mongolia. However, unlike a case of the North American Continent, neighboring country Mongolia is the land of the drying where there is no sea regrettably. Therefore, in the case of the China, the China should begin, after demanding the understanding and cooperation of the Mongolians, and then Japan should contribute to the construction of the fresh water suppling system as described in this text after ending the consultation between China and Mongolia. At present, China itself and some other countries imagine catching up with the United States in GDP and will overtake it for a while from before. However, I have stated that China did not have the Chinese ambition realized as far as China could not find the way where the Chinese should advance from the history that its own country walked on. As for this reason, simplicity is extremely clear. This is because fresh water resources are extremely poor to guarantee a rich life for the people equally in China with about 1/5 of the world's population. For this, I have showed approval to Three Gorges Dam Construction plan since 1990, but it was only redistribution of fresh water resources within a country and was never a plan to increase the fresh water resources. If the fresh water resources which are the most important to human are not economic growth being filled up, the disproportion may produce extremely serious international strains in the future.

Even if the 20th century began, natural science progressed, and the Industrial Revolution was over, the Chinese leader Mr. Mao Zedong had stepped forward to the way to major economic power as he could not notice the characteristic that its own country had. His ignorance produced starvers who were said to be tens of millions of people, and his fret that crossed out the history of his mistake conducted the stupid Cultural Revolution that had erased many racial living proofs. Strangely, it is still hard to understand the person who crossed out history, like an Islamic extremist, to me with object of respect of the national existence in the present age. Later many leaders of China were chased for the settlement of the man-made disaster that he left, and seemed to spend only an investigation of the exclusively at hand economic growth.

The result brought major economic power at the cost of the natural environments that both rivers and air are polluted. Furthermore, ignoring the ruined natural environments of its own country, it is not an intelligible thing that the Chinese measure to appeal to for the development of other countries by an AIIB plan. If the AIIB plan tells the regeneration of nature of its own country to want to borrow the power from other countries, it can be understood, and Japan should cooperate appropriately if it is a main purpose. It will contribute to the check the rise in China's sea surface to regenerate the continent nation which was full of green at ancient times. If China will stand up as a friend having the same cultural sphere for future humans, Japan should support even real fresh water supply in both finance and wisdom. Is such an action of Japan not good for the tomorrow's earth again ?

13. Addendum

In response to a contribution request, I worked out a design and began to make it into sentences, and now one year more passed. However, I have unfortunately come across some big events that I have to add about the state of the global community and about the matter of China which held the key role for tomorrow's peace of East Asia. Therefore, I cannot finish this article by ignoring these issues, because they are critically important to the status of human survival.

(A) The choice of Japan in response to a suffering EU facing refugees in a confused Islam zone

The most serious situation is the modern movement that a large number of refugees from the Islam society will flock in EU countries. As always, the Japanese government thinks about the maintenance of peace and order in Japan and is going to offer financial support to the concerned countries and/or organizations. Only money is not enough to settle present situation since the total number of refugees has reached about 4 million. I do not think that the population acceptance capacity in EU countries can bear the pressure of refugees from the confused Islam society. Even if it will take time to absorb the pressure of refugees, it will be to maintain the acceptance system of the refugees and emigrants, depending upon the conditions of each country throughout the world, and to help each EU country. Such a system must also contribute to the population movement with a climate change predicted in the future. This situation seems to become a good chance to push careful Japan to revise the current system for the acceptance of emigrants and refugees for fear of the aggravation of peace and order.

However, terrorists may be hidden in recent refugees so that it may give further anxiety to the Japanese for the acceptance of refugees and immigrants. But, in the other, Japan has already entered in the times of the population decline, and a decrease in working population becomes the urgent problem. Actually, if the birth rate in Japan remains the present value of around 1.34, it is a terrible story that we cannot estimate an ultimate drop of the population level in Japan. If the population decreases greatly, the extremely low self-sufficient ratios of food and energy supply that have been worried in Japan will become a problem to be solved by the following gradual decrease of population. Thus, the urgent policy problem may call on Japan to hurry and check the birthrate.

Nonetheless, the lack of work force in Japan is already serious and Technical Intern Training Program (TITP) has been founded in1993 for its solution. Here, TITP accepts the youth of developing countries as technical intern trainees, and after training them into journeymen in each field who can speak Japanese, it is a system that let them work there during some period of time. Here, the youths matured into journeymen while confirming a degree of achievement by taking a certification examination regularly, although this TITP has been often reviewed for various kinds of defects. Such a society situation suggests that a potential background in acknowledgement of the acceptance of refugees and emigrants exists in Japan.

However, almost Japanese are extremely careful about the acceptance of refugees and emigrants. The reason is because Japanese does not want them to disturb the world's safest country which was ranked by *Lifestyle9* in America the last year. As for this evaluation, I can understand from my own experience in which I have previously visited 23 developed countries for the attendance to societies and conferences. Generally, the aggravation of the peace and order in a small society is caused by a lack of communication among inhabitants, which comes from the inability to share a language in the area. From such circumstances, unless a naturalization or permanent residence applicant has any special skill, they do not have even procedures of the permission application on the Japanese government. Therefore, Japan has been said to the country which is strict with refugees and emigrants. As a matter of course, if people are in hope living in Japan on the occasion of refuge or emigration, it seems that they like not only the wonderful nature and culture of Japan but also appreciate the peaceful society of Japan. By the way, the communication between people is an indispensable element to maintain the good society in Japan. To that end, we cannot but demand some Japanese ability from a domiciliation applicant in Japan, the degree of which could read, hear and write Japanese at a level for elementary school 1 or 2 year.

In the history of humans, the race movement which is called to be the outbreak of refugees generated around the Islam area can be considered an offer of opportunity for interracial marriages and mixed blood off springs. As a result, the diversity of humans would spread with time and allow humans to adjust to the forthcoming unknown environments of global warming. Human migration on a large scale may be difficult after the drying up of fossil fuels. Japan should keep the outbreak of refugees in progress alive now as the 4th chance for interracial mix, following the 3rd chance interracial mix with the Chinese, producing Japanese children. This seems to raise the adaptation potential of Japanese to unknown environments in the future by increasing genetic diversity.

To that end, the Japanese government should change the acceptance system for foreigners choosing domiciliation in Japan hastily. For instance, the Japanese Embassy and Consulate of the world should provide opportunities for entrance examinations on Japanese ability twice a year to the domiciliation applicants in Japan. For entry, people 12 years or older, the Japanese government accepts domiciliation or naturalization of certified applicants who passed the Japanese examination promptly. For a family of 5 or more to qualify for entry, 2 people are required to have successfully passed the entrance exams. Although the entry of big families is not recognized to block the entry of terrorists. This policy of the Japanese government to open a window of acceptance to refugees and immigrants will be Japan's support to the EU countries, having trouble for the inflow of Islam refugees.

Recently on the night of this Nov. 24, NHK Japanese Broadcasting Cooperation reported a working population prediction by the Ministry of Health, Labor and Welfare of Japan. If the economic growth rate at present conditions remains at 2% for the next 15 years, the working population will likely decrease to 1.82million people by 2030, and if it continues remaining at 0%, it will fall to 0.79million, 12% as the rate of the total population. The aging Japanese population continues to advance. So even if the drop of the birth rate is prevented, judging from this situation, it may be said that the Japanese policy to open a window is greatly desired for supplementing a workforce drop, as well as maintaining peace and order with the refugees and immigrants from the Islam area.

Although being disappointed recently, a new trend does not like the acceptance of refugees from the Islam area in both EU and America. If the refugees of the Islam area do not show the posture to learn the local language for mutual understanding but persists in heterogeneous civilization, I can also understand that the existing nations are afraid of the aggravation on peace and order.

One of the conditions that human beings could live in a community was communication as mentioned in the 3rd item. Therefore, if the refugees from the Islam area intend to continue persisting in at least their original religion, they should try to learn the local language to adapt themselves to the countries where they want to immigrate. When people lack the means of mutual understanding, the peace of the community will not be kept, in which human beings could not live in a pre-existing society. It is also a reality that most Japanese feel uneasiness for their acceptance. To overcome this situation, I want them to show that they could be understood if the people from the Islam area communicate with the Japanese.

Anyway, the amplification of the genetic diversity which will gradually colonize within the Japanese must give a Japanese ability for a big adaptation to the violent change of the global environment. The active cooperation with the EU countries through the acceptance of refugee of Japan will be appreciated not only in the global community but also in Japan. If it comes to the amplification of the genetic diversity, by which the Japanese will obtain a greater ability for adaptation against severe climate change in the future.

(B) How China should act after the abolition of the One-child policy?

China announced the withdrawal of the One-Child Policy on Oct. 29. 2015. Here I want to consider the significance of the news to the neighboring countries. Although this policy led to the low birth rate in China premeditatedly as in the case of Japan, it was accompanied with various kinds of evils besides the social aging. Therefore, I want to push forward a story from the background why the Chinese government couldn't but draw up a policy that is regarded as human right violence.

The background that China could not but take the One-Child Policy is thought to be misleading of the leader of Chinese

Communist Party, Mr. Mao Zedong who is still evaluated as a founding further of the country. I cannot understand why his photograph is still raised in Tiananmen, because he has divided China into discriminative societies of the farm-village family and city family registers that limit a basic human right in 1958, in addition to the Great Leap Forward throughout 1958-1961 as already mentioned in the 5th item. In spite of them, he had further caused "Great Cultural Evolution" in 1966 as the leader of the communist party, which was planned for the recovery of his own honorary by agitating a youth group named "the Red Guard", who had destroyed historic and cultural heritages. Most members of the Chinese Communist Party who experienced such a bad system and stupid policies must have learned that the ability of food production does not increase in proportion to population growth under these conditions. As if it was waiting is his death in 1976, it would be natural to take on "the One-Child Policy" in 1978 as a population adjustment system appropriate to the increase of the food productivity.

Also the death of Mr. Mao gave an opportunity to escape from the economic activity that continued being delayed at the time. On this occasion, the person who played a leadership role was Mr. Den Xiaoping who had been forced to be a hermit by him. Though it is my own thought, would he pay attention to the total population of China increasing as for approximately 1.7 times, reaching 93.7 million, during a period of only 25 years of the past economic stagnation? Mr. Den who had studied in France in his youth and inspected Japan and America after the liberation from hermit life might have thought that the adaptation of a premediated population growth policy appropriate to the increase of the food-supplying power was indispensable to plan the growth of China's economy. I suppose that the new Chinese administration thus decided the adoption of the so-called "One-Child Policy" in 1979 for Chinese economic reform. This policy takes root in China as so-called "socialist market economy" and follows rapid economic growth, but the birth rate changes at a 1.2 level after 2002, and brings even a gentle drop of the population increase speed (China Statistical Yearbook 2013).

However, this One-Child Policy has given distortion in population structure in China like a decrease in working population in Japan. In addition, it was accompanied by the evils such as the birth of children at abnormal sex ratio and within family register, etc. These evils became more serious than a suppressant effect of the population expansion, and so the Chinese government came to cancel the One-Child Policy on Oct.29, which had been corrected for various kinds of defects many times. As far as, to here, it is happy that the discriminatory family registry system to downplay the most disgraceful human right will leave the abolition by 2020.

The abolition of the One-Child Policy will help maintenance of the work force in China, but it seems that the evil to come from the big difference in sex ratio of children continues being actualized for a while into the future, which may resultantly bring the situation that increases the number of young men who cannot find spouses in China.

In the times of the Japanese economy of the lost decade (1991-2000) and of the economic bubble (1986-1991), because the young women from the farm village had moved to the city, the young men on the farm couldn't but demand a partner either from South Korea or China which still remained in a lower economic growing stage. After both countries have accomplished economic growth together, the search for a bride had moved to the Vietnam and the Philippines where the economic level was low. In the present times when most countries in Asia are accomplishing economic growth, the border for encounters of youths seems to disappear probably because network environment has been regulated well. In the future 10-20 years after this, the young men of China is thought to have an extremely more difficult time seeking partners, because the economic growth has finished also at the neighboring countries. At this point, the withdrawal time of the One-Child Policy might be late even if the consent or 2 children may contribute to improvement of the sex ratio, but a negative effect to lose a valve for adjustment of the population increase appropriate to the increase of the food-supplying power adversely may invite a more serious situation in the future of China. In other words, even if the accepted delivery is only up to 2 persons, I cannot but estimate an increasing population in China in the long term even if the degree is very low.

The important thing is whether the Chinese government can control the increase of population in the range that can balance with the self-sufficient ratio of food under a predicted climate change. The problem is the reality that there is not a level of rain and snowfall in the whole area of China that can guarantee cultural civic life and high economic activities even in the present. When I watch what they say of Chinese people on the interest, I have often been reminded of their belief that China will develop into a major economic power of the world's best GDP exceeding the USA soon, and the Chinese will be able to live a rich life. However, it does not seem for me that the school in China tells children the fact that one of the conditions guaranteeing good life is to have a stable supply of fresh water. Will the Chinese communist party members be afraid to express the inability of Mr. Mao Zedong to teach the importance of the fresh water supplying capability? To teach the truth of historic events is vital in education.

Apart from light from the sun, children should have learned that the maximum resource for survival of humans is fresh water which continues existing semi-permanently on the earth. Further, the water is merely included in our body in approximately 60%, but goes in and out of approximately 1,300 ml in the human body in large quantities every day. The most important thing is that the substitution resources will be fresh water in the future when humans use the limited fossil fuels exhaustively. At that time, humans must live depending on electrons and hydrogen which are separated from fresh water. If water molecules could be separated cheaply into H^+ and electrons artificially, humans may open a way to transfer atmospheric CO_2 into organic matters as artificial foods. This suggests a possibility that agriculture could disappear and change into industry in the future. Even if the food production depends on the conventional natural environments at this point, hopefully it will not become extinct.

In spite of the IPCC report of old 2003 when global warming was not serious yet, the annual rainfall gross amount per person in China was only 2.293m^3, about 1/5 of America, and its level is equal to no developed country, being approximately 68% in comparison with Japan. At that time, nevertheless, the self-sufficient ratio of foods in China was almost filled, because the underground water was also useful. According to the report of the 5th IPCC, however, China is included in the zone where a decrease in rainfall would be predicted with global warming. Besides, it is a well-known story that the excessive use of the underground water in China has caused subsidence in the northwest land of China which may be concerned with the increasing frequency of earthquakes by bringing a distorted stratum in the Chinese inland, although some people explained it on the websites that it was due to the construction of the Three Gorges Dam that stores fresh water for the people in the northern land of China.

Besides, it is well-known story that the excessive use of the underground water in China causes subsidence in each domestic place. In recent years, on the other hand, the frequent occurrence of earthquakes is reported as news in connection with the construction of Three Gorges Dam which is built to store fresh water. A question is why the broad-based subsidence in the northwest area of China acts to increase the frequency by bringing a distorted stratum.

The underground water must dry up soon in the northwest area of China. Against expectations under global warming, the localization of rainfall may bring heavy rainfall also in the northern area of China. Even if this area shall be blessed with such a good fortune, however, the global warming does not seem to respond completely to fresh water demand there, because vast population greatly promotes the consumption of the water. The withdrawal of the One-Child Policy was done with disregard to such future risk. If it would be accompanied by a further increase of the population, China would be a country which will starve for fresh water in the future. Two changes of the policy, the revision of the Family Registration System and the withdrawal of the One-Child Policy by the Chinese government, should be evaluated fairly so more important human rights are protected by nations. In this situation, however, it is necessary that the Chinese have to face each other directly in the near future, until China becomes a country in which the population will be balanced with a supplying capacity of fresh water. To that end, that was the only way left for China to reduce population to proper size while raising the fresh water supplying capacity.

(C) The duty of Chinese occupying 1/5 of the human population

At last, the Chinese government came to show a fair original posture in deference to the human rights of the Chinese themselves, but it delayed the time of the population decline in China beginning since 2030 predicted in the World Population Prospect the 2015 Revision. In other words, these changes in policy will certainly hold a decreasing speed in working population, and will also contribute to improvement of the sex ratio between men and women. However, it was revealed that the population growth in China continued at least in year 2030 and later.

On the other hand, I have stated that in the northwest area in China the fresh water which is essential to human life is very short, which is thought to become more serious with progress of the global warming. According to FAO AQUAS TAT 2014, for example, the quantity of aquatic resources per person was only 2,127m^3/year in China, whereas those in America and Japan were 10,169m^3/y and 3,337m^3/y, respectively. The reason why the quantity of the aquatic resources is big in America is that it includes the polar region Alaska, whereas Japan has sleeping mountains so that most rainwater runs into the sea. However, China has neither the supplement of an extra aquatic resource nor the geographic feature. Therefore, China cannot but depend on the seawater conversion for the aquatic resource, in which people need a large quantity of energy while they also have trouble with the disposal of large amounts of inorganic salts included in the seawater. Maybe, China could not but depend on energy from fossil fuels for the conversion of seawater and for the industry to reuse the residual inorganic salts as resources while releasing CO_2 for the time being.

By the way, I cannot but think that the poorness of the fresh water supply capability in this situation may begin to cast a serious shadow over the future Chinese's economy. Speaking of which, the shortage of fresh water in China cannot be thought to be settled with increasing the conversion of sea water. On reading articles on "Science Portal China 2015" on Web site, I can see that the Chinese government works with various kinds of water conversion of seawater all over the country. However, the shortage of fresh water does not seem to be the level that can be settled by depending on the conversion. Not only it is predicted that the increase of the population continues for a long term, but also it is discovered that the ancient vegetation of the Chinese inland underground is unexploited with heavy oil and natural gas. As the result, the water demand in China further increases with more recent digging.

When one of the drought disasters in the continent inside is damage from salt breeze, it is not a good to use salt water for digging of the fossil fuel in the inland. Chinese people want to change the energy source from coal to natural gas which is unexploited under the desert, since coal releases PM2.5 in large quantities. They should wrestle in a dig of the natural gas using fresh water, if they think about the future environment in China. Therefore, it is apparent that there is a close relation between the stable and sustainable prosperity of the Chinese society in the future and the degree of fresh water demand.

Nevertheless, the Chinese prime minister Mr. Xi Jiping works hard at the construction of artificial islands for securing of gas field rights and interests, while bringing neighborhood countries' feeling of strain in defiance of international law. Already, China is the world's biggest CO_2 discharge country and the pictures in its capital Beijing show the situation that most cars cannot run without headlights even in the daytime because of air pollution by PM2.5. In COP21 in France of this early winter, he spoke about the bailout for warming checks of 70 billion dollars in a speech, and afterwards showed posture to participate in a frame of international coordination for the first time. The switch to his cooperation posture would change the American posture, that has hesitated about participation till then, and contributes of all of the world grew as the fruition of the frame which confronted a check of the global warming for the first time in this COP21. If he realizes the responsibility of the world's largest CO_2 discharging country, he should work on reproduction of nature itself in the interior land of China, and cancel artificial island construction that causes much of the rise of sea level, and try to relax the strain between China and the concerned countries. In this COP 21, there were cries for help from the small island countries that might be buried in seawater by a rise in sea level that collected international sympathy. It is sad thing for me, China coming out faces me, is running only for a right monopoly of the fossil fuel even if it ignores the international law without worrying about the submergence of the artificial islands by the rise in sea level.

I want to say the same thing once again. At first, those artificial islands of China want to submit the question about the legitimacy of the claim that it is the expulsion on its own country's territory. If it dated back to about 15,000 years, Malaysia, Vietnam, Philippine and Taiwan were the adjoining ground called a Sundaland continent and the existing South China Sea was an inland sea in the continent. The continent was the home of a similar Mongolian spot which also partly settled down in Japan. Soon the Han race went into Hainan Island at the south end of the Eurasia Continent, but a drop of the sea surface gradually began, and a Southeastern Area consisting of big and small islands was developed. Judging from Chinese ancient Chinese documents, both Chinese and Japanese ships seemed to navigate the South China Sea with the later state. In the early 12th century, the pirates called the Japanese Pirates (Wakou in Japanese) according to the ancient Chinese documents, had looted the Korean outskirts exclusively, but when the 15th century began, they had gone south and troubled the Chinese coast. However, the Japanese pirates went out from the end of the 15th century to the 16th century, and many merchants came to come and go as large-scale trader and opened the Japanese town in Vietnam and Thailand. Those historical structures are left in each country, but the prosperous trade became extinct by Japanese policy of seclusion before long. After Japan withdrew by the national policy, the dominion of control of the South China Sea moved to Ming and Qing dynasties from Japan, but based on these historical facts mentioned above, the dominion of the islands in the South China Sea should have been moved to Vietnam, Philippine, Taiwan and Malaysia according to current international lows of the sea.

Although China was able to identify such histories, in defiance of Vietnam, Thailand and Philippine, the Chinese government has begun to insist on the dominion of the archipelago there by the cause of the Sino Centrism, as soon as it is reported that there are fossil resources underneath the East

China Sea and the South China Sea. Unfortunately, their action has also produced a fear for China in the Japanese islands, Senkaku where the fishermen for tuna lived peacefully before war. It is a sad story that anyone in Chinese leadership can analyze exactly when the greatest resources for tomorrow's earth are not fossil fuels but fresh water. Fortunately, China owns a very large country where enormous undiscovered fossil resources are hidden. The biggest mission that any Chinese leader should do is to work on reproduction of nature of the own country that has been destroyed by Chinese ancient leaders.

In other words, the problem that a Chinese leader is assigned is to raise the fresh water self-sufficiency ratio in China even a little. Under these situations, not only is China the world's greatest CO_2 discharge country without enough supplying of fresh water but also the air in the China is polluted by PM2.5, the companies in China could not work normally and the Chinese dream for the greatest GDP over America could not be granted. According to IEA (2013), the energy self-sufficiency rate in both China and America remained to an 80% level, but most energy in China had been produced by burning the coal (70.4%) which was buried in large quantities at home. On the other hand, the heat in America had been severed mainly by oil (39.9%) and natural gas (25.3%). It would be natural to think that the serious air pollution in China is caused by the energy that is dependent excessively on coal. When I read Science Portal China (2015) on the web, by the way, I found news that there are gas fields of the greatest class in the inland of China in March and May of 2014, respectively.

If there is a possibility to reserve a large quantity of fossil fuel besides coal in China, it would not be necessary for the Chinese government to construct the artificial islands which may be buried in seawater with global warming while causing big friction internationally. I have understood the energy policy of the Chinese government to depend on natural gas rather than coal, in order to suppress the outbreak of PM2.5 troubling inside and outside the country.

According to UNCTAD2015, the import quantity of natural gas in China increases in late years and reaches 5th place. Furthermore, the Chinese government begins the laying of pipelines between neighboring countries to increase the import of natural gas. In addition, they increase the number of natural gas fields in the East China Sea by canceling regrettably a joint development agreement with the Japanese government one-sidedly. However, even if China changes the energy policy to natural gas dependence for PM2.5 discharge check, the speed of global warming does not decrease dramatically, which only hastens the submergence of the artificial islands. Will they forget fossil fuels are a limited resource and will dry up sometime soon?

If the Chinese government had wisdom, they should use both fund and technique for the reconstruction of its own country for the future, but not for the construction of the artificial islands. Preferably, I want Chinese leader, Mr. Xi Jinping, to become the first prime minister who will work on the reconstruction of the inland of its own country while the prices of oil and gas are relatively lower. If so, his name must remain in world history forever as the person who saved humans from the rise in sea level with global warming.

As described previously, China on the regeneration of the inland with little rainfall cannot but certainly demand support from Japan for fresh water supply. And this demand should give a chance to challenge the construction of a new Japan which could gather and store fresh water from rain and snowfall without draining it into the sea as much as possible. The favorable correspondence of Japan for the support request of the fresh water supply from China will show the way of one constructive cooperation in response to COP21 on the earth scale.

If this collaboration will be realized, it will be a great chance to give a large merit on the Japanese side as roughly described in item 13. I will add, China gives the opportunity for Japan to build the world's first network system for supplying fresh water in Japan, which would function and allow humans to live in the future environment of global warming. By the way, if the times come when we can use neither fossil fuels nor atomic energy for electric power supply, humans cannot but live depending upon natural energy and biomass. At that time, the Japanese cannot but utilize more of the territorial sea area ($43 \times 10^4 km^2$) which is larger than the small territorial land area ($38 \times 10^4 km^2$) to make use of light from the sun, the velocity of the wind, the wave force etc.as an electric energy source generally.

Therefore, if there is no support of the electric supply from China, Japan has to increase energy dependence upon the ocean.

So as to use the territorial sea area for electricity generation, we must destroy the marine environmental ecosystem that is a property of human joint ownership, which lets the coastal fishery of its own country weaken. In other words, Japan does not hurt the marine richness of East Asia so that the support of the electricity power supply from China is necessary, and it can contribute to the global community through the maintenance of the global environment.

Thus, the Japanese economy would develop with the completion of the electric power supplying network within the country, such as with the transportation network and with the communication network. Maybe, it would show the state of country under a global environment appearing after global warming by the completion of the freshwater-supplying network, leading the world. Although the base of the Japanese economy is on the rice production in the rice field, the freshwater-supplying network must contribute to prevention of drought disaster to be caused by meteorological localization which could be generated by global warming. Probably, the support request of the fresh water supply from China must lead to the construction of this network for fresh water supply unpreceded in the world of Japan. Moreover, the cooperative work between these two countries would make a successful model to produce a peaceful earth appropriate for human survival in the coming century.

14. Concluding Remarks

With this book, the author has talked before we drew the future image of the global environment, fossil fuels will dry up、and what people in future generations should do for their children's children. In addition to these, I would like to explain why the collaborative relationship between China and Japan is put in the title of this book.

By the way, it is apparent that the accumulation in the atmosphere of CO_2 which was brought by the economic activities of humans themselves pulling back the earth to ancient times when a mammoth had lived at the Arctic Ocean coast, and that the countries which had much exhausted CO_2 were located in the latitude zone in the Norther Hemisphere. On a happy note, however, in COP 21 all humankind agreed with The Paris Protocol, for restraint of the CO_2 emission, because it may delay the drying up of fossil fuels. Nevertheless, as far as humans live on earth, it is clear that they will use up the available fossil fuels and must change the energy source from fossil fuels to H_2 provided from water.

When I think like that, the area where the human being lives on the future earth will be limited to the place blessed with fresh water or to the place where fresh water exists at least. Therefore, if I look for an area blessed with the most rainfall in the Northern Hemisphere on the earth, the applicable place becomes the East Asian countries located in the west of the

Pacific, the steam from which is condensed, rains, and comes back to the earth continuous rotating. The area is called the Asian Monsoon Zone, China of the GDP world 2nd place and Japan of the world 3rd place are included there. These two countries have a similar cultural sphere that shared the same Chinese character (kanji) in the closest relationship since the ancient times. At modern times, however, Japan has disturbed the good relations. Still, it became like that after the war, and Japan had begun to walk the recurrence to those days again, based on a peace constitution. After China suddenly began to get involved in marine resources, however, the action of China brought feelings of strain in the East Asian area. Then, I judged it led to talking about a state of tomorrow's humanity which I write about bilateral coexistence. Thus, I came up with the title of the article. Finally, I itemize the point at issues here as a summary.

(I) An event epoch-making above all through an argument in the COP21 is to have born fruits in The Paris Protocol in which all human kinds make cooperating efforts to suppress the discharge of CO_2. This result must delay that time when the global warming actually causes pain to human, such as the subsidence of land by the rise in sea level and such as the difficulty of breathing under the increasing atmospheric concentrations of CO_2. As a result, The Paris Protocol might give us the hope that we could transfer in the H_2-dependent time before human using up the fossil fuels. If it becomes so, the human being may get preparations to shift to the H_2-society from cheap fossil fuels. In the time, however, a part of H_2 must be used as the materials of the organic matter composition. If so, it is supposed that O_2

accumulates in the atmosphere expressively, by which much organic matters are broken down oxidatively. The human must act sometime soon in order to suppress the rise in the atmospheric O_2 concentration, but The Paris Protocol gives the human time to devise it in order to evade excessive harmful action of O_2.

(II) In order to save human from the tragedy of the rise in sea level with the global warming, China as a nation of the GDP 2nd place should notice that it benefits in both the prosperity of the own country and world peace to make an effort for the reconstruction of nature in the Eurasian Continent, while getting the support of the fresh water supply from the rain and snow fall in comfortable Japan.

(III) Japan has fresh water resources, but it is under the ocean environment and demands electricity to acquire H_2 on the territorial sea when the H_2 society arrives. Therefore, Japan should supply the fresh water which is dripped by the snow and rainfall to China to expect reproduction of the Eurasian Continent as far as it is possible, if there was a support request from neighboring country China with responsibility to humans. A part of the quantity of water supply should demand return from China as electricity produced in the Chinese continent inside.

(IV) Even if the drying land spreads by global warming, humans will become able to change the earth of the drying into forest or farm, if humans can quench the drying ground with fresh water. Resultantly, this possibility may gain population capacity of the earth dramatically. Or, even if humans can quench only a part of the place of frying place with fresh water, humans should be able to let the industry to revolutionize agriculture in cooperation with the communication-oriented society through the years.

Furthermore, if it is the time when humans can artificially composes DNAs of necessary enzymes, humans may invite the times when artificial reduction of atmospheric CO_2 and the following production of macromolecules can provide various kinds of foods to humans. At that time, humans can achieve the industrialization of agriculture, finish the premediated food production system, and may completely break into the H_2 society independent from the fossil fuel.

By the 1st Industrial Revolution the start of the steam engine in the 18th century, humans have begun to build a rich society. With the development of the electricity industry in the 19th century, the 2nd Industrial Revolution had occurred. Then, the 3rd Industrial Revolution successively assumes information with a computer in the 20th century. What humans should aim at the 21st century is to regard it as "century of water" or "century of forest". This may be the 4th Industrial Revolution to be called the industrialization of agriculture. As described in Item 8 (E), one of the possibilities that may be realized soon is the advanced nano-cellulose industries started in Japan. These industries using plant fibers must be gradually replaced by the current chemical companies using fossil fuels. In the times when people will begin to see that the energy depends mainly upon the sun, plastic, rubber, plants etc. which tie their fate to fresh water. The sun picks up H_2 from fresh water, and the plants cannot live in a little amount of fresh water.

If it shall be able to arrive early in such times, humans may not use up fossil resources. Humans may not hold the concentration of atmospheric CO_2 by taking atmospheric CO_2 in foods, medicines and various organic matters freely by using H^+ obtained from H_2O, because people will be able to compose artificially all kinds of DNA with which they will be able to compose freely all kinds of necessary enzyme proteins. At that time, people may realize the complete recycling society of the C element that the atmospheric CO_2 concentration shall be kept at a constant level by the spread of the artificial composition industry for various organic matters. Therefore, there will be another homework left unfinished for survival of humans who took for CO_2 absorbent to reduce the level of the superabundant atmospheric O_2 in future air. Although the evolution of the creatures on the earth was adaptation to increasing atmospheric O_2 concentration from 0.2% to 21%, I do not think that humans themselves will be able to adapt at a super oxidative state of the atmosphere. Unfortunately, most of the inorganic elements which have become the guard of O_2 such as Fe or Ca etc., which existed abundantly on primitive earth, has been already oxidized.

2016/1/11 Former FAO Consultant
Honor Prof. of Tohoku Univ. Dr. Yohji Esashi

Afterword

I have spent a long time, approximately 2 years on the completion of this book. This small booklet is created with the appeal from a director of some overseas publishing company since the autumn of 2014. I thought there must be some reason for an old man like me to receive such an invitation. Probably it was thought whether my action on the occasion at the first Earth Summit held a quarter of a century ago, in 1992, in which I canceled the original bill submitted by the United Nations side and I played a leadership role. The original bill could not endure the issues with the progression of global warming.

Although disasters and abnormal weather continue to occur in the world, unfortunately, there was no international agreement to hold down global warming which I worried about. Until now, the agreement has been controlled by China which was world's greatest CO_2 discharge country. Even after finishing COP20 in 2014, because China had 2nd place in world GDP, the world refused to participate in international action for suppressing global warming by giving priority to the profit of its own country and by placing oneself under the pretense of developing countries. I am concerned about this deplorable situation and decided to respond on the proposal of the publishing company.

At this time, the main title of my book which I conveyed to the director was "Mutually Reciprocal Roads for Fresh Water and Electricity in the Sea between Japan and China", though the same as this in the subtitle, when I told him that I had an idea of writing a series to a magazine. Such a writing came to mind from fear to the supremacy principle of the Chinese government that left domestic environmental pollution and brought feelings of strain throughout East Asia. Particularly, it has given an excuse for revision of the peace constitution in Japan. Therefore, I wanted to demand action for good sense maintenance of peace in Asia from the Chinese who occupy 1/5 of the human population.

Because I obtained the Chief Editor's consent, I have begun to send my manuscript at every good point of the limit to him, but before long, I was advised about the publication of this book, not a series. Because my writing will include many severe indications for China, I doubted whether the big publishing company could be valid in commercialism. Could it undertake publication and would it not lose a large market called China? Even in the 1990 Tiananmen Square protest incident when trade regulations had been enforced globally, only the American products such as tobacco etc. were continued being sold in Beijing because of the greater existence of China as a market.

During the period of writing, moreover, the Chinese government canceled the one-child policy which was one symbol of human right suppression and went along with the Paris protocol in COP21. These events were the matters that might reduce the significance of my existing book.

As being supposed, although I sent all my manuscript, changed the title for this book with the tomato photograph, and got permission from the chief editor this spring, his aggressive posture was not felt to some extent.

And the manuscript had been left without reason which was later revealed after a brief while. However, I was able to guess from former experience according to commercialism. Therefore, the conclusion was to investigate a method to have a Chinese in particular read my intention without being poisoned by commercialism. Then, I told Mr. E. Ohuchi running the local publishing company Honnomori in which there was a friendly acquaintance with me until now. Immediately, he understood to look by some means or other,

My motive to write this book was because the fate of tomorrow's earth must be entrusted to the Chinese who produced the very large drying area in the central part of the Eurasian Continent in ancient times and who occupy 1/5 of the world population now. For regeneration of the large wasteland, the awareness of Chinese own at first is necessary, as well as, a great deal of fresh water is necessary, which is absolutely short in supply.

Accordingly, Japan must help them by supplying fresh water in large quantities, which will as a result contribute to minimizing a Japan own damage due to the rising of sea level in Japan. On the other hand, many countries will follow the action of China, Thus, the firm mutual two-way street will be established between Japan and China. In Japan, it will probably be an opportunity for the construction and maintenance of the network for fresh water supply, which resultantly the big recession refraining from the end of the Tokyo Olympic Games will become an illusion. In contrast, China will have the opportunity with the condition with world's greatest GDP exceeding the United States by being able to serve fresh water demand. The large GDP countries, China 2nd and Japan 3rd must carry out whiffler, only hand in hand together, for humans to coexist peacefully in the global environment, prepared for the warming that will come.

As a result, if the Chinese government would really demand a grant of fresh water supply from Japan for the regeneration of its own country, it will promote the construction of mutual two-ways to premise equivalent exchange by returning electricity to tomorrow's Japan. The offer from China must give help after holding the Tokyo Olympic Games in Japan when it will be unable to look for a way for the next development. Maybe, the request of China will let Japan play a pioneer position of the purpose that constructs the network of aqueducts which are built around a main road for running fresh water through the Japanese islands and parts of the main road should extend to China through the bottom of sea. On the contrary, Japan may keep neighboring marine ecosystems in good condition, if we could be able to make use of the electricity sent from China in return. These events must be a big dream for a new country to prepare for global warming in Japan which may defeat nothing to be possible after the Tokyo Olympic Games. In China, the regeneration of the Eurasian Continent would begin and would help humanity survive possibly after the advancement of global warming. It may be said that it is a source of eternal peace in East Asia

2016/12/24

Dr. Yohji Esashi (1933～)

Honor Professor of Tohoku University, Sendai, Japan.

Bachelor's degree from Tohoku University in 1955, and M.A. from Tohoku University in 1957. He was awarded a Ph.D. in Science from Tohoku University in 1961.

Afterwards, he was an Assistant Professor of the Institution of Agricultural Science at Tohoku University from 1961 until 1971, during which he held a postdoctoral post with Professor A.C. Leopold at Purdue University in the United States from 1969 until 1971.

In 1972, he became an associate Professor of the Department of Biological Science at Tohoku University, and then was promoted to Professor in 1974.

In 1993, he became the first Chairman of the new Environmental Biology section in the Faculty of Science of Tohoku University, which was started in the association with Botanical Garden affiliated with the University, and retired in 1996.

He served on the committee as a Plant consultant for FAO to establish the Svalbard International Seedbank for seed storage from 1990, and pre-discussed during the Earth Summit held in Brazil in 1992. He was head of the Japan Wood Seed Research Institute, from 1997 until 2009.

Born in Sendai, Miyagi, Japan.